T0331688

Soft
Nanomaterials

World Scientific Series in Nanoscience and Nanotechnology*

ISSN: 2301-301X

The Series aims to cover the new and evolving fields that cover nanoscience and nanotechnology. Each volume will cover completely a subfield, which will span materials, applications, and devices.

Published

For further details, please visit: http://www.worldscientific.com/series/wssnn

(Continued at the end of the book)

Volume
19

World Scientific Series in
Nanoscience and Nanotechnology

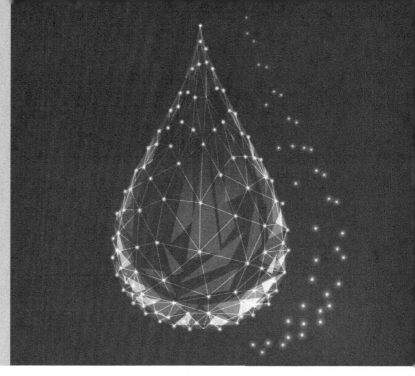

Soft
Nanomaterials

Editors

Ye Zhang
Okinawa Institute of Science and Technology, Japan

Bing Xu
Brandeis University, USA

 **World Scientific**

NEW JERSEY · LONDON · SINGAPORE · BEIJING · SHANGHAI · HONG KONG · TAIPEI · CHENNAI · TOKYO

Published by

World Scientific Publishing Co. Pte. Ltd.

5 Toh Tuck Link, Singapore 596224

USA office: 27 Warren Street, Suite 401-402, Hackensack, NJ 07601

UK office: 57 Shelton Street, Covent Garden, London WC2H 9HE

British Library Cataloguing-in-Publication Data
A catalogue record for this book is available from the British Library.

World Scientific Series in Nanoscience and Nanotechnology — Vol. 19
SOFT NANOMATERIALS

Copyright © 2020 by World Scientific Publishing Co. Pte. Ltd.

ISBN 978-981-120-102-8

For any available supplementary material, please visit
https://www.worldscientific.com/worldscibooks/10.1142/11300#t=suppl

Desk Editor: Rhaimie Wahap

Typeset by Stallion Press
Email: enquiries@stallionpress.com

Printed in Singapore

Preface

As a prominent feature of life, soft nanomaterials perform sophisticated functions and promise diverse applications; as an area of research, soft nanomaterials has a broad scope. This small collection of researches of leading experts in the field, undoubtedly, also reflects these two aspects. This book underscores the key feature of soft nanomaterials, illustrates the essential strategies for making them, and highlights their promises.

In Chapter 1, Cui *et al.* illustrated how to engineer "self-assembling supramolecular nanostructures for drug delivery". For the purpose of decreasing overall toxicity and enhancing therapeutic efficiency in cancer therapy, self-assembly of peptides or block copolymers generates supramolecular nanostructures that target cancer cells by ligand-receptor interactions, enzymatic reaction, or photochemistry. This chapter covered the state-of-art topics and examples, as well as remaining challenges for drug delivery. In Chapter 2, Ryu *et al.* focused on "bioactive supramolecular assembly of peptide amphiphiles to control cellular fate". After reviewing several representative examples of peptide amphiphile for self-assembly or enzyme-catalyzed self-assembly, the authors introduced the applications of the assemblies, from tissue engineering to controlling cellular functions. In addition to describe the assemblies of peptide amphiphiles for antibacterial applications and drug delivery, this chapter showed the promise of spatiotemporal control over the self-assembly of molecules in cellular environment. In Chapter 3, Yamanaka summarized the progress the works on "supramolecular gel electrophoresis of protein". After giving a brief history of electrophoresis, the authors discussed how to develop a low molecular weight hydrogelator leads to supramolecular gel electrophoresis (SUGE). This chapter is a fine example of practical applications of soft nanomaterials. In Chapter 4, Nie and Yi summarized "self-assembly of polymer-grafted inorganic nanoparticles into functional hybrid materials". After detailing the design, synthesis, characterization of polymer-grafted inorganic nanoparticles, the authors showed the elegant and diverse superstructures made by the self-assembly of these "hair" inorganic nanoparticles (HINP) at different conditions. This chapter is an excellent starting point for developing more elaborate soft nanomaterials for more sophisticated functions. In Chapter 5, Xing and Yang also showed the power of surface-modification for making soft nanomaterials by describing "surface coated NIR light-responsive nanostructures for imaging and therapeutic applications". The authors summarized the surface engineering of several well-established nanomaterials, gold nanoparticles, upconversion nanoparticles, carbon nanotubes, and graphene for applications in drug delivery, photothermal therapy, and photodynamic therapy. In Chapter 6, Maruyama briefly summarized "surface functionalization through polymer segregation", which illustrated the complicate and dynamic features of polymer coating on solid substrates. In Chapter 7, Reif *et al.* discussed "nucleic acid hairpins: a robust and powerful motif for molecular Devices". After introduction the uniqueness of hairpins, the authors illustrated the power of DNA hairpins in nucleic acid circuits, locomotion, and autocatalysis. This chapter also highlighted future challenges for *in vivo* applications and for using self-replication to produce large DNA nanostructures.

As shown in this book, self-assembly is one of the simplest and the most explored approach to make soft nanomaterials from a wide range of building blocks. We hope readers who are interested in the research of soft nanomaterials will be inspired by these pioneering works. The field of soft nanomaterials is emerging as one most active research frontiers. We expect that many exciting discoveries and promising applications are to come from innovative exploration of soft nanomaterials. We also like to thank all the authors for their excellent contributions.

Ye Zhang

Okinawa Institute of Science & Technology, Japan

Bing Xu

Brandeis University, USA

Contents

CHAPTER 1

Self-Assembling Supramolecular Nanostructures for Drug Delivery

MICHAEL PORTER,[*] **RAN LIN,**[*,‡]
MAYA MONROE[*,‡] **and HONGGANG CUI**[*,†,‡,§]

[*]Department of Chemical and Biomolecular Engineering,
The Johns Hopkins University, USA
[†]Department of Materials Science and Engineering,
The Johns Hopkins University, USA
[‡]Institute for NanoBioTechnology, The Johns Hopkins University,
3400 N Charles Street, Baltimore, USA

1.1. Introduction

Humans have used medicinal concoctions for thousands of years to improve the quality and longevity of life. Over the past several decades, the use of molecular therapeutic agents has risen drastically, particularly in the realm of cancer diagnosis and treatment. However, with the increasing time and expense required for new drug development, this interest has started to fade. Attention has begun to shift from the synthesis and discovery of new chemical entities (NCEs)

[§]Corresponding author: hcui6@jhu.edu

to the design of new formulations and delivery systems for existing therapeutics that achieve improved clinical outcomes. Despite this shift, the low patient survival rates, invasive procedures, high costs, and imprecise nature of mainstream cancer treatments leave much to be desired. Supramolecular nanomedicine has begun to gain prominence within the field of drug delivery due to its relatively low-cost and simplicity. "Supramolecular" refers to the various intermolecular forces that arise from small scale interactions, while "nanomedicine" describes the length scale at which these interactions occur. In contrast to the mainstream methods that utilize free drugs with relatively narrow therapeutic indexes, nanocarrier delivery systems use the unique properties of nanoscale objects to decrease overall toxicity and enhance therapeutic efficiency.

To attain successful cancer treatment, the drug delivery vehicle must be engineered to circumvent the numerous physiological, extracellular, and intracellula barriers that exist to protect the body's cells. Self-assembling supramolecular nanostructures have attracted significant interest in the field of drug delivery due to their tunable pharmacokinetic (PK) profile and drug targeting specificity. To perform their required function, cancer drugs must surpass the body's defense mechanisms, enter the desired site of action, and accumulate to an appreciable concentration.[1,2] Most cancer treatments are administered intravenously to maximize bioavailability. However, this exclusively relies on the free drug's PK profile, often resulting in low target specificity and undesirable side effects due to off-site accumulation. With a discrete nanocarrier, the design of the drug-carrier system can be tailored to the desired target tissues, protecting the cargo from and allowing it to penetrate physiological barriers.

In addition to their targeting capabilities, nanocarriers improve the solubility of the cargo, enhance its PK profile, and optimize its therapeutic index. Ideally, drug carriers should demonstrate a high drug loading capacity and efficiency to maximize the delivered drug dose, thereby limiting any resistance cells may develop to the drug. Tuning the PK profile and cellular uptake require optimization of particle size, charge, and shape. In general, positively charged, spherical nanoparticles must have a minimum diameter of 8 nm to avoid immediate renal clearance. However, the optimal size and shape for uptake depend on the desired therapeutic effect and cellular targets.[3–7]

The rational design of self-assembling nanostructures requires an understanding of the unique physical, chemical, and biological properties that arise at these length scales. Self-assembly occurs through noncovalent interactions between the molecular building blocks. These interactions include hydrophobic, electrostatic, and hydrogen bond interactions and can be manipulated through molecular design, environmental stimulus, and implementation of co-assembling molecules.[8] The interactions between each subunit produce a well-ordered nanostructure that can be used as an effective drug carrier. These resulting metastable structures only exist above a critical assembly concentration (CAC) and are often engineered to dissociate in the presence of a particular biological or chemical stimulus.[9,10] The subunits can conform and adapt to each other during assembly, resulting in more internal order.

Although self-assembled nanostructures hold great promise, few products have obtained FDA approval. Even those treatments that have received approval fail to demonstrate significantly improved antitumor efficacy over their free drug counterparts. Doxil®, a formulation of liposome encapsulated doxorubicin (DOX), an anthracycline antibiotic, was approved by the FDA in 1999 for the treatment of ovarian cancer. Clinical trials showed a decrease in cardiotoxicity compared to free doxorubicin, but the two formulations exhibit very little difference in patient survival rates.[11] Similarly, Abraxane®, an albumin-based, paclitaxel nanoparticle approved by the FDA in 2005, has a response rate of only 21% despite being the most effective metastatic breast cancer treatment on the market.[11] This is likely due its lack of an active targeting mechanism and weak PK profile.[11] These therapeutic formulations are demonstrably inadequate, leading many to seek further improvement of nanomedicine. In this chapter, we will explore recent developments in self-assembling nanostructures and evaluate them in the context of improved drug delivery efficiency for cancer treatment.

1.2. Engineering a Drug Delivery System

Designing an effective drug carrier requires an understanding of the intricate biology of the human body and tumor microenvironment. In addition to enhancing passive tumor targeting, supramolecular nanostructures serve to protect the drug from the body's defense mechanisms and improve its overall PK profile. Lone drug molecules are highly susceptible to renal clearance due to their relatively small size; any particle with a diameter less than 8 nm is subject to rapid removal by the kidneys.[7] Furthermore, the neutral charge and hydrophobicity of most anticancer therapeutics make them prime targets for drug metabolism.[12,13] During this process, the kidney and liver chemically convert such drugs into polar substances that can be more easily cleared.[12,13] Self-assembled nanostructures circumvent these issues by increasing the relative size of the drug and engulfing it in a protective stealth layer.

Conversely, there is an upper size limit that must be considered in nanoparticle design. The reticuloendothelial system (RES, or mononuclear phagocyte system) comprises the host's foreign body immune response. Primarily composed of macrophages, the RES relies on phagocytosis and opsonization of foreign matter in the liver, lungs, spleen, bone marrow, and lymph nodes.[14] Extensive RES uptake may result in asymptomatic sequestration and storage of the drug in these organs, leading to potential organ enlargement and dysfunction. Nanoparticles that exist above 200 nm are found to be quickly eliminated through the RES, thereby implementing another design limitation on drug delivery vehicles.

Following administration, the cancer drug must navigate through additional obstacles before reaching the tumor tissue. If taken orally, the drug carrier must overcome the first pass effect, which is the initial elimination of a foreign substance through intestinal and hepatic degradation.[15] Alternative routes of administration include intravenous, percutaneous, intramuscular, or intrathecal injection and inhalation. Each route displays unique advantages and disadvantages contingent on the mechanism of drug delivery and the intended destination of the supramolecular drug. The mode of administration can drastically alter the PK profile, changing the apparent concentration of the drug in the bloodstream.

Regardless of its route of administration, the drug carrier must ultimately make its way into the cancerous cell. The uncontrolled proliferation of tumors results in the formation of uneven vasculature with a larger-than-normal pore size (>100 nm as compared to

5–10 nm) within the cancerous tissue.[16] Additionally, tumor tissue exhibits poor lymphatic drainage, leading to particulate buildup in the interstitial space. This phenomenon, aptly named the enhanced permeation and retention (EPR) effect, can be exploited by nanoparticle design to improve passive drug uptake and retention (Fig. 1.1).[17] Nanoparticles with diameters between 150–200 nm have been found to benefit from the EPR effect, demonstrating improved tumor drug delivery efficiency.[18,19] Combining passive targeting via the EPR effect with active tumor-specific targeting mechanisms involving ligand or antibody-modification can significantly enhance the tumor specificity and anticancer efficacy of the nanoparticle drug delivery system.

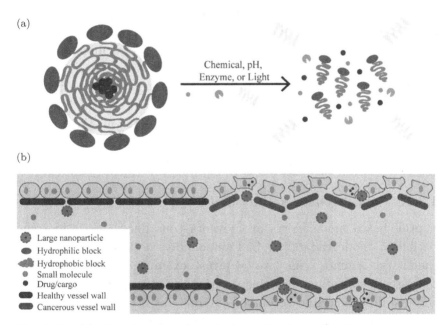

Fig. 1.1. (a) Nanostructures (spherical micelle pictured) self-assemble under physiological conditions due to the amphiphilic nature of their building blocks. In the presence of a chemical, pH, enzyme, or light trigger, the supramolecular structure can disassemble to release its cargo. (b) Tumor architecture is often characterized by uneven and leaky vasculature and poor lymphatic drainage, enabling the accumulation and retainment of larger or otherwise impermeable molecules within the cancerous tissue. The EPR effect is exploited extensively in nanoparticle design to achieve passive tumor targeting.

1.3. The Building Blocks of Self-assembly

Supramolecular nanomedicine offers a high degree of customization, enabling it to overcome various physiological barriers. Strategies include unique targeting mechanisms to circumvent nonspecific uptake and shielding to protect the therapeutic cargo from nonspecific binding, adsorption, and elimination. One common way self-assembled structures implement the shielding strategy is to encapsulate a small hydrophobic drug within an outer hydrophilic layer, thereby protecting the drug from unfavorable interactions. These structures exist in thermodynamic equilibrium as meta-stable objects, subject to conformational changes based both on the surrounding environment and the chemical structures of their components.[20]

Supramolecular structures derive their function from the chemical composition of their unimers, which are the individual self-assembling components of the structure. Manipulating the shape and function of these structures requires a fundamental understanding of their chemistry. The base components that can be used for self-assembly are nearly endless, including polymers, peptides, and RNA/DNA scaffolds.[8,21–23] Most unimers are amphiphiles that derive their specific properties from the components that make up the subunit. We will further discuss the advantages and limitations of two specific unimer classes, peptides and block copolymers.

1.3.1. *Peptide based self-assembly*

Peptide-based nanocarriers are advantageous due to their biocompatibility, biodegradability, biofunctionality, and relatively simple chemistry. Amino acids are a prime example of nature-driven, self-assembled design: proteins derive their structure and function through intramolecular secondary interactions between amino acids. Peptides are built from 20 naturally occurring standard amino acids, each with unique properties arising from their hydrophobicity, charge, polarity, and size. The specific combination and sequence of an amino acid chain drastically changes the overall properties of the resultant peptide. In turn, the secondary structures of a peptide, such as α-helical and β-sheet complexes, display unique chemical and structural properties that can be exploited for drug delivery.

Most anticancer drugs are small and hydrophobic and can be functionalized via chemical conjugation onto a specific peptide sequence to form an amphiphilic prodrug. This one-component nanomedicine strategy of direct conjugation of the drug onto the unimer allows for the precise control and fine tuning of drug loading.[24] Cheetham *et al.* developed a drug-peptide amphiphile capable of assembling into either nanotube or nanofiber morphologies depending on the drug loading capacity. Camptothecin, a DNA-topoisomerase I inhibitor, was conjugated onto a hydrophilic, β-sheet forming peptide sequence through a reducible disulfybutyrate linker attached to a cysteine residue. This peptide-drug amphiphile exhibits a drug loading capacity between 23–38%.[25] The exact number of camptothecin molecules per peptide was controlled and varied using the amino acid lysine as a branching point for the attachment of multiple cysteines for drug conjugation. In a reducing environment, such as the cell cytoplasm, the nanofibers demonstrate steady release kinetics and sufficient cell toxicity *in vitro*. The vast yet facile customization offered by peptide unimers allows for chemical specificity ideal for tumor-targeted drug delivery.

MacKay *et al.* synthesized $\sim$100 nm peptide nanoparticles for doxorubicin (DOX) delivery (Fig. 1.2). In this system the hydrophilic portion is comprised of an elastin like peptide while the hydrophobic segment contains DOX molecules conjugated to cysteine residues via pH-sensitive hydrazone bonds. Although the nanoparticle is relatively stable at a pH of 7.4, it demonstrates tunable, first order release kinetics at pH 5.0.[26] As is the case with most peptide-based structures, this particle consists of biocompatible and degradable components and is relatively easy to synthesize with a high yield. In the context of chemotherapeutic drug delivery, the particle size is ideal for the utilization of the EPR effect and avoidance of renal clearance and RES uptake. These characteristics, along with the monodispersity and tunable release kinetics of the particles, are important design criteria to adhere to when engineering a supramolecular nanostructure for drug delivery.

Peptide-based structures offer a wide variety of properties, conformations, and functions due to the nearly unlimited number of peptide sequences. This approach is relatively inexpensive, the chemistry

(a)

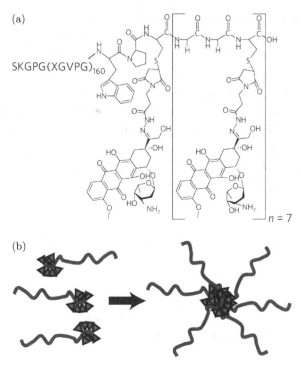

SKGPG(XGVPG)$_{160}$

(b)

Fig. 1.2. (a) Elastin-like peptides are joined with a block of cysteine residues, to which the DOX cargo is conjugated through hydrazone bonds. (b) A micelle structure self-assembles through the sequestration of the drug loaded hydrophobic regions (red triangles) at the center and the hydrophilic peptide sequence (blue) at the periphery.[26] Adapted from Ref. 26. (Copyright © 2009, Nature Publishing Group)

is simple, and it facilitates the incorporation of additional factors such as cell adhesion and signaling sequences.[27–30] However, most current research utilizes relatively short sequences that produce three dimensional structures via intermolecular interactions. With the aid of computational tools, it may be possible to explore more complex three-dimensional structures that arise from a single peptide strand, similar to current studies involving DNA and RNA assembly.[31,32]

1.3.2. *Block copolymer self-assembly*

The self-assembling capabilities of block copolymers come from their multiple sections, or "blocks". Two, three, or more different polymer segments can be linearly conjugated onto one another to create diblock, triblock, and multiblock copolymers, respectively. The blocks can also be conjugated to one another in a nonlinear fashion, resulting in branched copolymers. Relatively recent improvements in polymer synthesis allow for the precise control of molecular weight, block size, and branching, thereby making the production of these diverse block copolymers feasible.[30] This structural customization can be used to tune the mechanical, morphological, and chemical properties of the nanostructure.

Block copolymers derive their unique structural and functional properties from the sequence and characteristics of their subunits. The properties of these block copolymers must make for an ideal supramolecular material for drug delivery, notably with regards to their biocompatibility and biodegradability. Like peptide unimers, these block copolymers are amphiphilic in nature. Each section, or "block", consists of a single type of polymer that is either hydrophilic or hydrophobic. The polymer used for each block can be interchanged or altered as needed to provide a variety of different functionalities, including the incorporation of polymers conjugated to different targeting ligands or polymers with sidechains that provide secondary interactions. Both composition and length of each block dictate nanostructure morphology, which varies drastically and can take the form of micelles, nanospheres, polymersomes, and filaments.[33–37] The high degree of customization afforded by block copolymer assembly is due to the variety of polymers available that display sensitivity to pH, temperature, and other environmental factors.

Polymers that contain charged functional groups are commonly used as components in self-assembling block copolymers. Poly(β-amino ester), a hydrophobic polymer, displays sensitivity to its surrounding pH. This property depends on polymer pK_b, which varies with molecular weight; making pH sensitivity tunable via degree of polymerization.[38] Ko *et al.* exploited this property to design a pH responsive block copolymer from poly(β-amino ester) and hydrophilic methyl ether PEG (mPEG). The block copolymer was synthesized using Michael-type polymerization to ensure a pK_b of approximately 6.5. Above this pH the copolymer forms stable micelles, below it the micelles fall apart, thereby enabling its nanostructures to utilize the acidic pH of the tumoral interstitial space to trigger increased drug release. The resulting micelle demonstrated a DOX loading capacity of 74.5% within the hydrophobic core. More importantly, the micelles exhibited improved antitumor efficacy as a consequence of their rapid drug release in tumor tissue due to their pH sensitive demicellization behavior.[39]

Hyaluronic acid (HA), a naturally occurring polymer found in the extracellular matrix, has also been utilized in nanocarrier design. Ganesh *et al.* exploited HA's ability to bind to CD44, a receptor that is commonly overexpressed by cancer cells. The near 100% encapsulation efficiency of therapeutic siRNA in their branched, diblock copolymer highlights the potential for HA-PEG/PEI systems.[40] This system also demonstrates that the target specificity of block copolymers makes them a useful tool in nanoscale drug delivery.

The incorporation of additional blocks allows for increased customization. Sun *et al.* developed a triblock copolymer featuring mPEG, hydrophobic poly(ε-carprolactone) (PCL), and cationic poly(2-aminoethyl ethylene phosphate) (PPEEA) to create a three-layer, nanoscale micelle for therapeutic siRNA delivery. The hydrophilic mPEG and PPEEA blocks comprise the outer shell; mPEG improves the immunogenicity and circulation time and PPEEA serves to bind and release the siRNA from the nanoparticle. The hydrophobicity of PCL allows for the formation of the micelle structure and furthers thermodynamic stability under physiological conditions.[41]

Yang *et al.* developed a more complicated multiblock copolymer, three-layer, vesicular system for the pH-sensitive delivery of encapsulated DOX and superparamagnetic iron oxide (SPIO) NPs (Fig. 1.3). By utilizing various molecular weights of folate (FA) (A), PEG (B), and poly(glutamate hydrazone DOX) (C), Yang *et al.* constructed an A-B-C-B-A-B-C-B-A multiblock copolymer with a drug loading capacity of 14 wt% and 46 wt% for DOX and the SPIO NPs, respectively. Acceptable stability was achieved by crosslinking the inner PEG layer of the vesicle. PEG and FA were used in the outer layer to promote non-immunogenicity and tumor targeting, respectively.[42] Although the inclusion of various polymer, therapeutic, and diagnostic agents in the system limited its maximum drug loading capacity,

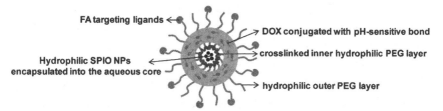

FA targeting ligands

DOX conjugated with pH-sensitive bond

Hydrophilic SPIO NPs
encapsulated into the aqueous core

crosslinked inner hydrophilic PEG layer

hydrophilic outer PEG layer

Fig. 1.3. Schematic of Yang's multiblock copolymer vesicle design. FA targeting ligands and PEG encircle DOX and SPIO NPs in the hydrophobic and hydrophilic sections of the vesicle, respectively.[42] (Copyright © 2010 American Chemical Society).

their incorporation demonstrates the versatility and utility of block copolymers for both cancer drug delivery and imaging applications.

1.4. Tumor Targeting and Drug Release

In the realm of cancer treatments, self-assembling materials have demonstrated improved performance due to their reduced toxicity, high drug loading capacities, and increased therapeutic indexes. Circulation time and target specificity are improved through the encapsulation of hydrophobic drugs, which serves to prevent premature enzymatic, chemical, pH, or hydrolytic degradation. Despite this protection, the drug must still reach its site of action to perform its intended function. In addition to overcoming the aforementioned barriers to drug delivery, the nanostructure must destabilize and release the drug in the presence of its biological target for the drug to exert its pharmacological activity. With that in mind, one of the most critical features in the design of a drug delivery system is the mechanism by which the therapeutic cargo will be released.[30]

1.4.1. *Ligand targeting and enzymatic release*

Enzymes, proteins, surface receptors, and various other chemical components are critical in moderating the metabolic activity of a cell. Tumor cells are often characterized by the overexpression of a single receptor or enzyme, which can be used for receptor-mediated tumor targeting or enzyme-responsive drug release.[43]

The resistance of tumor cells to multidrug and radiation therapies is central to the challenge of targeting and eliminating them.

This resistance is the result of their overexpression of glutathione (GSH), a tripeptide reducing agent found in abundance within the cancer cells.[44] Koo and coworkers exploited this anomaly in their triblock copolymer nanoparticle design. PEG-b-poly(L-lysine) (PLL)-b-poly(L-phenylalanine) (PPha) micelles were stabilized by crosslinking the free amines of the PLL block using a bifunctional, GSH-sensitive 3,3'-dithiobis(sulfosuccinimidylpropionate) linker.[45] Wang *et al.* synthesized a reversible block copolymer micelle from PCL-b-polyphosphoester (PPE)-b-PEG. PPE was modified with a crosslinking thiol group, stabilizing the micelle and preventing the release of the DOX payload. Both labs utilized thiol-based linkers that are cleaved in the presence of GSH in the design of their nanostructures. This results in both tumor targeted release of the loaded drug and GSH depletion in the tumor tissue.

Alternatively, cysteine Cathepsins (Cat), a group of lysosomal proteases, are also highly upregulated within cancer cells. Cathepsins are active in cell death and proliferation; specifically, CatB regulates protein turnover and is overexpressed in several tumor types including prostate, colon, and breast cancers.[47] Lock *et al.* developed a molecular probe that utilizes Förster resonance energy transfer (FRET) between fluorescent DOX and a quenching molecule. The drug is conjugated onto a cell penetrating peptide sequence through a CatB specific peptide linker, resulting in an amphiphilic beacon molecule. In the presence of CatB, the peptide linker is cleaved to release free DOX capable of fluorescence and therapeutic activity (Fig. 1.4). This vehicle protects DOX from premature degradation with the burial of the CatB substrate/linker in the self-assembled micelle structure.[46] The beacon's sensitivity to CatB and its self-assembling capability increase the efficiency and specificity of DOX delivery to the target cancer cells.

Similarly, Andresen *et al.* targeted secretory phospholipase A_2 (sPLA2), a cancer specific biomarker, using a phospholipid-based liposomal drug carrier. sPLA2 cleaves β-fatty acids to mediate lipid storage and has been found to play an integral role in cell proliferation and differentiation. Furthermore, sPLA2 is overexpressed by more than an order of magnitude in tumor cells, making it an ideal biomarker for drug delivery vehicles.[48] Anderson's non-hydrolyzable,

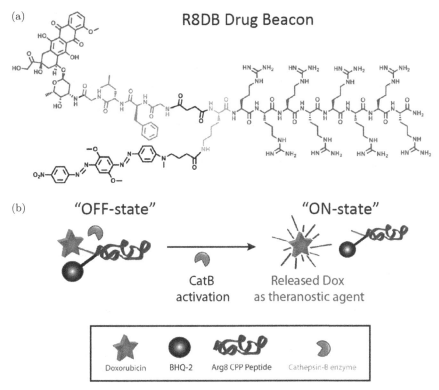

Fig. 1.4. (a) The structure of the drug beacon is shown, with DOX (red) conjugated onto the R8 cell penetrating peptide (purple) and BHQ-2 quenching molecule (black) through a CatB sensitive linker (green). (b) In the presence of CatB, DOX is cleaved to reveal a fluorescence signal.[46] (Copyright © 2015 American Chemical Society)

1-*O*-phospholipid based liposomes demonstrate significant stability under physiological conditions but succumb to the cleavage potential of sPLA2 to trigger the release of their cargo. Their drug delivery potential was explored using edelfosine (ED), which is known for its strong hemolytic properties. *In vitro*, liposomal ED exhibited significantly lower hemotoxicity in stark contrast to the otherwise low toxic dosage of free ED.[49]

Recently, reversible enzyme triggers have been explored in the context of self-assembly and degradation for more tunable drug release. Webber *et al.* studied the utility of a peptide amphiphile containing a substrate sequence for protein kinase A (PKA), a well

characterized cancer biomarker involved in cellular metabolism. In
the absence of PKA, the peptides display a fiber like morphology.
A serine residue is phosphorylated in the presence of PKA, resulting
in the disassembly of the nanofiber structures. The specificity and
reversibility demonstrated by this system offer more precise thermodynamic control of DOX release in the presence of cancer cells that
overexpress PKA.[50] These studies conducted by Lock, Andresen, and
Webber demonstrate the great promise of introducing mechanisms
for the safe and precise delivery of cancer drugs.

1.4.2. *pH-sensitive release*

The ubiquitous nature of pH makes it a powerful tool in targeted
drug delivery. pH gradients exist on multiple physiological levels; for
example, the human gastrointestinal tract and stomach demonstrate
a pH as low as 1.7, while late endosomes maintain a pH of about
5–6 through the use of proton pumps.[51,52] Drug delivery applications
for cancer treatment can exploit the abnormal pH found in tumor
tissue. Normal tissue is reported to have a pH of approximately
7.4 but some cancerous tissue has been reported to have a pH of
6.2–7.4.[53,54] This discrepancy, which is likely a result of the tumor's
high metabolic rate coupled with its limited blood supply and lymphatic drainage, has been used to facilitate the specific targeting of
tumor cells.

Wu *et al.* developed a pH responsive micelle using both mPEG-
b-poly(β-amino ester) (PAE) and AP-PEG-b-PLA copolymers (AP
is a tumor targeting peptide). The tertiary amine groups of PAE
have a pK_b of 6.5, which contributes to delayed demicellization and
drug release at a pH range of 6.4–6.8 (Fig. 1.5). Passive accumulation in tumor tissue via the EPR effect and active targeting and
prolonged tumor retention due to the presence of the AP peptide
enable the micelles to remain at the target site long enough for the
structures to destabilize in the acidic tumor microenvironment, effectively releasing their cargo.[55] In a similar study, Lee *et al.* synthesized
polymeric micelles composed of poly(L-histidine) (pHis)-b-PEG and
mixed micelles that also contained poly(l-lactic acid) (PLLA)-b-PEG
to tune the pH-sensitive drug release. The polyHis/PEG micelles

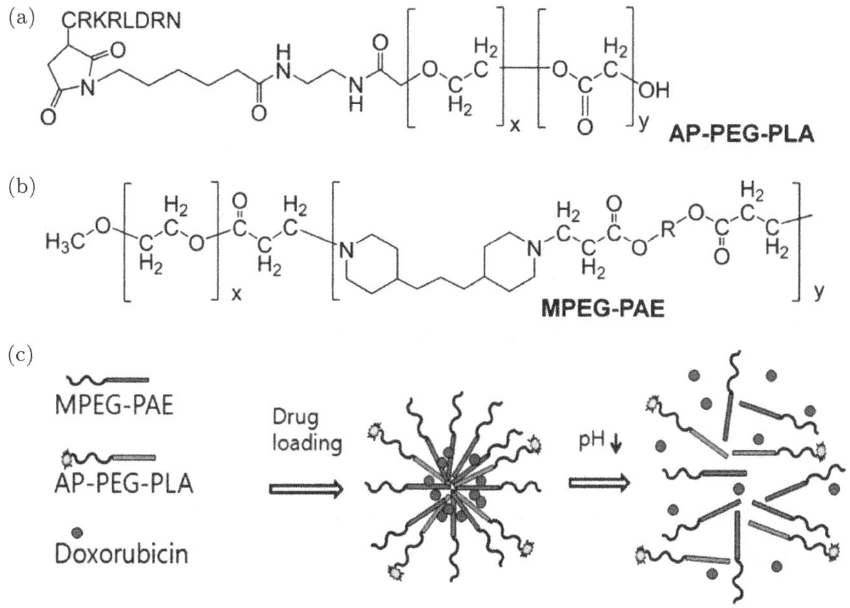

Fig. 1.5. (a) Chemical structures of AP-PEG-PLA and (b) MPEG-PAE. (c) DOX is released through the structural destabilization of the micelle by a pH drop, such as that experienced in the tumor microenvironment.[55] Adapted from Ref. 55. (Copyright © 2010 American Chemical Society)

demonstrated accelerated drug release starting at a pH of 8.0 with a further gradual increase in release rate until a pH of 6.8. The mixed micelles exhibited the accelerated drug release transition within the more acidic pH range of 7.2–6.6, making them more suited for tumor responsive drug release. To achieve further improved antitumor efficacy, the mixed micelles were modified with folic acid to enhance tumor cell internalization.[56]

Although pH-triggered release exhibits promise for cancer therapy, certain limitations restrict this method's *in vivo* potential. While tumor extracellular pH is slightly acidic, it is also heterogeneous and its relatively small pH range overlaps with that of normal tissue, decreasing the specificity of the drug delivery system. However, it is possible to pair this method with another preexisting targeting mechanism to circumvent these issues and create a more effective tumor targeting drug delivery system.

1.4.3. *Light-sensitive release*

The techniques mentioned up to this point have utilized physiological triggers for drug delivery. In contrast to these methods, light sensitive drug delivery systems respond to electromagnetic radiation to achieve a similar feat. Radiation at or below a wavelength of 700 nm can only penetrate through approximately 1 cm of tissue, limiting its application to treatment of cancer in the skin or superficial organs. Near-infrared (NIR) radiation with a wavelength up to 900 nm can reach deeper tissues and organs.[57] Utilizing the electromagnetic spectrum holds promise for an innocuous, tunable, and precise cancer drug delivery system.

Sun *et al.* designed a dendritic polymer based amphiphile that demonstrates ultraviolet (UV) and NIR sensitive drug release. Click chemistry was used to conjugate a diazonapthoquinone (DNQ) decorated poly(amido amine) dendron to PEG to form self-assembling amphiphiles. Demicellization occurs following the transformation of the hydrophobic DNQ core to a hydrophilic molecule in the presence of UV or NIR radiation.[58] These results suggest that a self-assembled, irreversible, light sensitive drug release mechanism could be beneficial for cancer treatment.

In a similar approach, Bondurant *et al.* developed a liposomal delivery capsule. PEG-liposomes were prepared from different ratios of PEGylated lipids, cholesterol, photosensitive lipid bis-sorbPC, and DOPC to calibrate the photosensitivity and drug loading capacity of the structure. In the presence of UV light, bis-sorbPC photocrosslinks with other molecules of bis-sorbPC. This simultaneously increases the permeability of the liposome and decreases its structural integrity, allowing the encapsulated drug to permeate through the bilayer and into the environment.[59] Although such methods hold promise for light triggered drug release, this technology remains elusive in medical contexts due to the potential health complications from radiation. Further research is needed to explore the potential of more benign photo activation methods.

Structural Morphology and Integrity

One of the challenges of designing a supramolecular drug delivery system is ensuring its robustness. The nanostructure must maintain its structural integrity prior to its arrival at the site of action to avoid the premature release or degradation of the cargo. After administration, self-assembled drug carriers experience dilution and are especially susceptible to harsh physiological conditions such as pH gradients, temperature, enzymatic degradation, and the body's immune defense and metabolic mechanisms (RES, first pass effect, etc.). Supramolecular nanostructures must be designed to withstand these conditions and still be capable of disassembling and releasing their cargo or revealing their drug linker in the presence of a target signal like the hypoxic tumor environment or an overexpressed enzyme. Regardless of its ultimate purpose, the architecture and structural integrity of the nanostructure are crucial to its function.

Altering the length and composition of the hydrophobic/ hydrophilic blocks of the amphiphile unimers is a direct method of tuning the stability and architecture of the self-assembled nanostructure.[60,61] The hydrophobic, ionic, and steric interactions that govern self-assembly are inherent in these characteristics and can be quantified through the critical packing parameter (p) as follows:

$$p = \frac{v}{l_c a_0}$$

Where v is the volume of the hydrophobic tail group, l_c is the critical chain length, and a_0 is the optimal group area of the hydrophilic head (Fig. 1.6).[62] If p is less than a third, the amphiphiles will self-assemble into spherical micelles. As this value increases, the predicted nanostructures are cylindrical micelles (nanotubes, $p = 1/3 - 1/2$), vesicles ($p = 1/2 - 1$), planar bilayers ($p = 1$), and inverse micelles ($p > 1$).

This parameter suggests that altering the physical components of an amphiphile chain can greatly affect the architectural outcome of self-assembly. Although inherently obvious, this parameter closely follows empirical findings. For example, Sugihara *et al.* reported on a poly(2-hydroxypropyl methacrylate) (PHPMA)-b-poly(2-(methacryloyloxy) ethylphosphorylcholine) (PMPC) block copolymer and the effect of increasing hydrophobic PHPMA block length (thereby increasing the value of the volume of the hydrophobic tail group, v, in relation to l_c and a_0) on the self-assembled morphology; inducing an architectural change from spherical, to rod shaped,

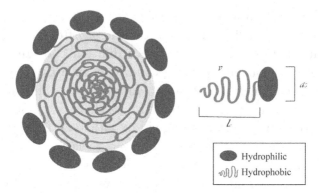

Fig. 1.6. The structure of a spherical micelle is comprised of many polymer subunits with embedded hydrophobic (blue) and outer hydrophilic (red) components. The critical packing parameter is calculated using the volume of the hydrophobic tail group (v), the critical chain length (l_c), and the optimal hydrophilic head group area (a_0).

to vesicular.[63] Zhang *et al.* came to a similar conclusion using a polystyrene (PS) and poly(acrylic acid) (PAA) block copolymer.[64] As the optimal head group area (a_0) of the amphiphile was decreased, the self-assembled morphology progressed from micelles to nanotubes to vesicles.

Dong *et al.* explored the limitations of these steric repulsions and entropic penalties by constructing a micelle from amphiphilic helix peptides conjugated to PEG. The stability of the micelle can be tuned by changing the location of the PEG chains along the length of the peptide amphiphile. As the PEG chain moves from the center of the micelle to the outer shell the geometric packing improves and the entropic repulsion alleviates, stabilizing the micelle.[65] This ability of simple design characteristics to effectively alter the stability of the nanostructure is consistent with the theoretical critical packing parameter.

The critical packing parameter provides a theoretical guideline for morphology and stability, but it cannot account for all the intricacies of self-assembly. Although most self-assembled unimers are amphiphiles, the exact composition and properties of these segments, such as their ability to form hydrogen or ionic bonds, are crucial in defining their interactions. For example, Paramonov *et al.*

investigated the effect of hydrogen bond forming amino acids in peptide amphiphiles and discovered that the order, number, and orientation of the hydrogen bond donor and receptor residues are critical in determining the morphology and stability of the nanostructure.[66]

Some researchers have utilized chemical crosslinking following self-assembly to reversibly anchor the individual subunits to one another. Stupp and coworkers applied this technique to a peptide-based nanofiber by incorporating four cysteine residues capable of disulfide bond formation in the unimer sequence.[67] Below a pH of 4, the peptides self-assemble into nanofibers, but in basic conditions the amphiphilic nature of the peptide unimer is altered and the nanofibers fall apart. Stability can be conferred by oxidizing the thiol groups of the cysteines following self-assembly to form disulfide crosslinks between the unimers of the nanofibers, allowing the fibers to remain intact at a pH of 8. These crosslinks are reversible and can be reduced to enable the disassembly of the structures. With a critical micelle concentration (CMC) of 0.01 mg/ml, the disulfide bridges increase the structural integrity and robustness of the micelle-like structures. Above a unimer concentration of 2.5 mg/ml, the fibers entangle into a self-supporting gel. Such hydrogels exhibit a unique potential for local drug delivery and are capable of encapsulating large biologics such as proteins. Their high viscosity and low mobility make them ideal candidates for local delivery, allowing the therapeutic to bypass several of the physiological barriers previously mentioned and be directly released at the site of action.[68] They also display promise for local delivery with systemic release due to their high loading capacity, well-controlled release, high stability, and environmental responsiveness, although this application is relatively unexplored.[68]

A similarly constructed triblock copolymer micelle for systemic delivery was developed by the Wang group using PCL-b-poly (2,4-dinitrophenylthioethyl ethylene phosphate) (PPE)-b-PEG.[69] PPE contains functional groups capable of forming disulfide bonds following oxidation, thereby stabilizing the micelles. A significant fraction of drug release occurred after reduction severed the crosslinks. These reversible systems may be precursors to a more robust and translatable nanostructure for *in vivo* cancer therapy and detection.

1.5. The Future of Nanomedicine and Its Clinical Translation

Supramolecular nanostructures have gained significant interest in the past several decades due to their potential for cancer treatment. Their high degree of customizability enables researchers to tailor their design for improved passive and active transport to the desired site of action. As discussed in this chapter, peptides, block copolymers, and various other molecular units have been used as the effective building blocks of self-assembled nanostructures. Additional chemical modification is often used to introduce specific targeting mechanisms, such as ligand conjugation for enhanced target cell internalization, the inclusion of special linkers or structures for enzyme, pH or light responsive drug release, or the addition of functional groups for stability and protection.

Despite their promise, very few nanomedicines are in clinical use due to inconsistencies between animal models and human trials, tumor heterogeneity, and other aforementioned challenges encountered by these drug delivery systems.[19] As an example, the EPR effect, one of the underlying principles behind passive targeting in cancer drug delivery, is not reliable in clinical trials. While the phenomenon is observed in certain animal models, it is not as pronounced in the human body.[70,71] Other barriers to clinical translation include safety concerns,[72] regulatory hurdles,[73] and an overall lack of reproducibility, large scale production, and trust in academic research.[74] Nanomedicine still has many issues to address before it is translatable from the lab to the clinic.

Self-assembling supramolecular nanostructures have developed a formidable presence in the field of cancer drug delivery. The topics and examples covered in this chapter represent the accumulation of knowledge thus far and the opportunities and challenges to be explored and overcome in this field. Although a great deal of progress needs to be made in academic and industrial research, the need for more potent, safe, and economical cancer therapies will continue to drive interest in this field.

References

1. Davis, M. E., Chen, Z. and Chin, D. M. (2008). Nanoparticle therapeutics: An emerging treatment modality for cancer, *Nat. Rev. Drug Discov.* **7**, 771–782.
2. Park, J., Fong, P. M., Lu, J., Russell, K. S., Booth, C. J., Saltzman, W. M. and Fahmy, T. M. (2009). PEGylated PLGA nanoparticles for the improved delivery of doxorubicin, *Nanomedicine* **5**(4), 410–418.
3. Choi, H. S., Liu, W., Misra, P., Eiichi, T., Zimmer, J. P., Ipe, B. I., Bawendi, M. G. and Frangioni, J. V. (2007). Renal clearance of quantum dots, *Nat. Biotechnol.* **25**(10), 1165–1170.
4. Alexis, F., Pridgen, E., Molnar, L. K. and Farokhzad, O. C. (2008). Factors affecting the clearance and biodistribution of polymeric nanoparticles, *Mol. Pharm.* **5**(4), 505–515.
5. Geng, Y., Dalhaimer, P., Cai, S., Tsai, R., Tewari, M., Minko, T. and Discher, D. E. (2007). Shape effects of filaments versus spherical particles in flow and drug delivery, *Nat. Nanotechnol.* **2**(4), 249–255.
6. Gratton, S. E. A., Ropp, P. A., Pohlhaus, P. D., Luft, J. C., Madden, V. J., Napier, M. E. and DeSimone, J. M. (2008). The effect of particle design on cellular internalization pathways, *Proc. Natl. Acad. Sci. U.S.A.* **105**(33), 11613–11618.
7. Liu, J., Yu, M., Zhou, C. and Zheng, J. (2013). Renal clearable inorganic nanoparticles: A new frontier of bionanotechnology, *Mater. Today* **16**(12), 477–486.
8. Cui, H., Webber, M. J. and Stupp, S. I. (2010). Self-assembly of peptide amphiphiles: From molecules to nanostructures to biomaterials, *Biopolymers* **94**(1), 1–18.
9. Chen, Z., Zhang, P., Cheetham, A. G., Moon, J. H., Moxley, J. W., Lin, Y.-a. and Cui, H. (2014). Controlled release of free doxorubicin from peptide–drug conjugates by drug loading, *J. Control. Release* **191**(10), 123–130.
10. Lomas, H., Canton, I., MacNeil, S., Du, J., Armes, S. P., Ryan, A. J., Lewis, A. L. and Battaglia, G. (2007). Biomimetic pH sensitive polymersomes for efficient DNA encapsulation and delivery, *Adv. Mater.* **19**(23), 4238–4243.
11. Dawidczyk, C. M., Kim, C., Park, J. H., Russell, L. M., Lee, K. H., Pomper, M. G. and Searson, P. C. (2014). State-of-the-art in design rules for drug delivery platforms: Lessons from FDA-approved nanomedicines, *J. Control. Release* **187**, 133–144.
12. Longmire, M., Choyke, P. L. and Kobayashi, H. (2008). Clearance properties of Nano-sized particles and molecules as imaging agents: Considerations and caveats, *Nanomedicine (London, England)* **3**(5), 703–717.
13. Takakura, Y. and Hashida, M. (1993). Macromolecular carrier systems for targeted drug delivery: Pharmacokinetic considerations on biodistribution, *Pharm. Res.* **13**(6), 820–831.
14. Saba, T. M. (1970). Physiology and physiopathology of the reticuloendothelial system, *Arch. Intern. Med.* **126**(6), 1031–1052.

15. Gibaldi, M., Boyes, R. N. and Feldman, S. (1971). Influence of first-pass effect on availablility of drugs on oral administration, *J. Pharm. Sci.* **60**(9), 1338–1340.

16. Wang, A. Z., Langer, R. and Farokhzad, O. C. (2012). Nanoparticle delivery of cancer drugs, *Annu. Rev. Med.* **63**, 185–198.

17. Prabhakar, U., Maeda, H., Jain, R. K., Sevick-Muraca, E. M., Zamboni, W., Farokhzad, O. C., Barry, S. T., Gabizon, A., Grodzinski, P. and Blakey, D. C. (2013). Challenges and key considerations of the enhanced permeability and retention effect for nanomedicine drug delivery in oncology, *Cancer Res.* **73**(8), 2412–2417.

18. Brigger, I., Dubernet, C. and Couvreur, P. (2012). Nanoparticles in cancer therapy and diagnosis, *Adv. Drug Deliv. Rev.* **64**, 24–36.

19. Bae, Y. H. and Park, K. (2011). Targeted drug delivery to tumors: Myths, reality and possibility, *J. Control. Release* **153**(3), 198–205.

20. Whitesides, G. M. and Grzybowski, B. (2002). Self-assembly at all scales, *Science* **295**(5564), 2418–2421.

21. Klok, H.-A. and Lecommandoux, S. (2001). Supramolecular materials via block copolymer self-assembly, *Adv. Mater.* **13**(16), 1217–1229.

22. Douglas, S. M., Dietz, H., Liedl, T., Hogberg, B., Graf, F. and Shih, W. M. (2009). Self-assembly of DNA into nanoscale three-dimensional shapes, *Nature* **459**(7245), 414–418.

23. Guo, P. (2005). RNA nanotechnology: Engineering, assembly and applications in detection, gene delivery and therapy, *J. Nanosci. Nanotechnol.* **5**(12), 1964–1982.

24. Su, H., Koo, J. M. and Cui, H. (2015). One-component nanomedicine, *Journal of Controlled Release* **219**, 383–395.

25. Cheetham, A. G., Zhang, P., Lin, Y.-a., Lock, L. L. and Cui, H. (2013). Supramolecular nanostructures formed by anticancer drug assembly, *J. Am. Chem. Soc.* **135**(8), 2907–2910.

26. MacKay, J. A., Chen, M., McDaniel, J. R., Liu, W., Simnick, A. J. and Chilkoti, A. (2009). Self-assembling chimeric polypeptide–doxorubicin conjugate nanoparticles that abolish tumours after a single injection, *Nat. Mater.* **8**(12), 993–999.

27. Zhang, X.-X., Eden, H. S. and Chen, X. (2012). Peptides in cancer nanomedicine: Drug carriers, targeting ligands and protease substrates, *J. Control. Release* **159**(1), 2–13.

28. Accardo, A., Salsano, G., Morisco, A., Aurilio, M., Parisi, A., Maione, F., Cicala, C., Tesauro, D., Aloj, L., De Rosa, G. and Morelli, G. (2012). Peptide-modified liposomes for selective targeting of bombesin receptors over-expressed by cancer cells: A potential theranostic agent, *Int. J. Nanomedicine* **7**, 2007–2017.

29. Sugahara, K. N., Teesalu, T., Karmali, P. P., Kotamraju, V. R., Agemy, L., Greenwald, D. R. and Ruoslahti, E. (2010). Co-administration of a tumor-penetrating peptide enhances the efficacy of cancer drugs, *Science* **328**(5981), 1031–1035.

30. Cheetham, A., Chakroun, R., Ma, W. and Cui, H. (2017). Self-assembling prodrugs, *Chem. Soc. Rev.* **46**, 6638–6663.

31. Rothemund, P. W. K. (2006). Folding DNA to create nanoscale shapes and patterns, *Nature* **440**(7082), 297–302.

32. Colombo, G., Soto, P. and Gazit, E. (2007). Peptide self-assembly at the nanoscale: A challenging target for computational and experimental biotechnology, *Trends Biotechnol.* **25**(5), 211–218.

33. Burt, H. M., Zhang, X., Toleikis, P., Embree, L. and William, H. L. (1999). Development of copolymers of poly(D,L-lactide) and methoxypolyethylene glycol as micellar carriers of paclitaxel, *Colloids Surf. B* **16**, 161–171.

34. Discher, D. E. and Eisenberg, A. (2002). Polymer vesicles, *Science* **297**(5583), 967–973.

35. Yoo, H. S. and Park, T. G. (2001). Biodegradable polymeric micelles composed of doxorubicin conjugated PLGA-PEG block copolymer, *J. Control. Release* **70**, 63–70.

36. Gref, R., Minamitake, Y., Peracchia, M. T., Trubetskoy, V., Torchilin, V. and Langer, R. (1994). Biodegradable long-circulating polymeric nanospheres, *Science* **263**(5153), 1600–1603.

37. Qian, J., Zhang, M., Manners, I. and Winnik, M. A. (2010). Nanofiber micelles from the self-assembly of block copolymers, *Trends Biotechnol.* **28**(2), 84–92.

38. Kim, M. S., Lee, D. D., Choi, E.-K., Park, H.-J. and Kim, J.-S. (2005). Modulation of poly(β-amino ester) pH-sensitive polymers by molecular weight control, *Macromol. Res.* **13**(2), 147–151.

39. Ko, J., Park, K., Kim, Y.-S., Kim, M. S., Han, J. K., Kim, K., Park, R.-W., Kim, I.-S., Song, H. K., Lee, D. S. and Kwon, I. C. (2007). Tumoral acidic extracellular pH targeting of pH-responsive MPEG-poly(β-amino ester) block copolymer micelles for cancer therapy, *J. Control. Release* **123**, 109–115.

40. Ganesh, S., Iyer, A. K., Morrissey, D. V. and Amiji, M. M. (2013). Hyaluronic acid based self-assembling nanosystems for CD44 target mediated siRNA delivery to solid tumors, *Biomaterials* **34**(13), 3489–3502.

41. Sun, T.-M., Du, J.-Z., Yan, L.-F., Mao, H.-Q. and Wang, J. (2008). Self-assembled biodegradable micellar nanoparticles of amphiphilic and cationic block copolymer for siRNA delivery, *Biomaterials* **29**, 4348–4355.

42. Yang, X., Grailer, J. J., Rowland, I. J., Javadi, A., Hurley, S. A., Matson, V. Z., Steeber, D. A. and Gong, S. (2010). Multifunctional stable and pH-responsive polymer vesicles formed by heterofunctional triblock copolymer for targeted anticancer drug delivery and ultrasensitive MR imaging, *ACS Nano* **4**(11), 6805–6817.

43. Andresen, T. L., Jensen, S. S. and Jorgensen, K. (2005). Advanced strategies in liposomal cancer therapy: Problems and prospects of active and tumor specific drug release, *Prog. Lipid. Res.* **44**(1), 68–97.

44. Estrela, J. M., Ortega, A. and Obrador, E. (2006). Glutathione in cancer biology and therapy, *Crit. Rev. Clin. Lab. Sci.* **43**(2), 143–181.

45. Koo, A. N., Lee, H. J., Kim, S. E., Chang, J. H., Park, C., Kim, C., Park, J. H. and Lee, S. C. (2008). Disulfide-cross-linked PEG-poly(amino acid)s copolymer micelles for glutathione-mediated intracellular drug delivery, *J. Chem. Soc. Chem. Commun.* **48**, 6570–6572.

46. Lock, L. L., Tang, Z., Keith, D., Reyes, C. and Cui, H. (2015). Enzyme-specific doxorubicin drug beacon as drug-resistant, *ACS Macro Lett.* **4**(5), 552–555.

47. Sloane, B. F. and Mohamed, M. M. (2006). Cysteine cathepsins: Multifunctional, *Nat. Rev. Cancer* **6**(10), 764–775.

48. Rillema, J. A., Osmialowski, E. C. and Linebaugh, B. E. (1980). Phospholipase A 2 activity in 9, 10-dimethyl-1, 2-benzanthracene-induced mammary tumors of rats, *BBA-Lipid Lipid Met.* **617**(1), 150–155.

49. Andresen, T. L., Davidsen, J., Begtrup, M., Mouritsen, O. G. and Jorgensen, K. (2004). Enzymatic release of antitumor ether lipids by specific phospholipase A2 activation of liposome-forming prodrugs, *J. Med. Chem.* **47**(7), 1694–1703.

50. Webber, M. J., Newcomb, C. J., Bitton, R. and Stupp, S. I. (2011). Switching of self-assembly in a peptide nanostructure with a specific enzyme, *Soft Matter* **7**(20), 9655–9672.

51. Dressman, J. B., Berardi, R. R., Dermentzoglou, L. C., Russell, T. L., Schmaltz, S. P., Barnett, J. L. and Jarvenpaa, K. M. (1990). Upper gastrointestinal (GI) pH in young, healthy men and women, *Pharm. Res.* **7**(7), 756–761.

52. Dominska, M. and Dykxhoorn, D. M. (2010). Breaking down the barriers: siRNA delivery and endosome escape, *J. Cell Sci.* **123**(8), 1183–1189.

53. Gerweck, L. E. (1998). Tumor pH: Implications for treatment and novel drug design, *Semin. Radiat. Oncol.* **8**(3), 176–182.

54. Vaupel, P., Kallinowski, F. and Okunieff, P. (1989). Blood flow, oxygen and nutrient supply, and metabolic microenvironment of human tumors: A review, *Cancer Res.* **49**(23), 6449–6465.

55. Wu, X. L., Kim, J. H., Koo, H., Bae, S. M., Shin, H., Kim, M. S., Lee, B.-H., Park, R.-W., Kim, I.-S., Choi, K., Kwon, I. C., Kim, K. and Lee, D. S. (2010). Tumor-targeting peptide conjugated pH-responsive micelles as a potential drug carrier for cancer therapy, *Bioconjugate Chem.* **21**(2), 208–213.

56. Lee, E. S., Na, K. and Bae, Y. H. (2003). Polymeric micelle for tumor pH and folate-mediated targeting, *J. Control. Release* **91**(1), 103–113.

57. Alvarez-Lorenzo, C., Bromberg, L. and Concheiro, A. (2009). Light-sensitive intelligent drug delivery systems, *J. Photochem. Photobiol.* **85**(4), 848–860.

58. Sun, L., Zhu, B., Su, Y. and Dong, C.-M. (2014). Light-responsive linear-dendritic amphiphiles and their nanomedicines for NIR-triggered drug release, *Polym. Chem.* **5**, 1605–1613.

59. Bondurant, B., Mueller, A. and O'Brien, D. F. (2001). Photoinitiated destabilization of sterically stabilized liposomes, *Biochim. Biophys.* **1511**, 113–122.

60. Mai, Y. and Eisenberg, A. (2012). Self-assembly of block copolymers, *Chem. Soc. Rev.* **41**(18), 5969–5985.

61. Hamley, I. W. (2011). Self-assembly of amphiphilic peptides, *Soft Matter* **7**(9), 4122–4138.

62. Khalil, R. A. and Zarari, A.-H. A. (2014). Theoretical estimation of the critical packing parameter of amphiphilic self-assembled aggregates, *Appl. Surf. Sci.* **318**, 85–89.

63. Sugihara, S., Blanazs, A., Armes, S. P., Ryan, A. J. and Lewis, A. L. (2011). Aqueous dispersion polymerization: A new paradigm for in situ block copolymer self-assembly in concentrated solution, *J. Am. Chem. Soc.* **133**(39), 15707–15713.

64. Zhang, L. and Eisenberg, A. (1995). Multiple morphologies of "crew-cut" aggregates of polystyrene-b-poly(acrylic acid) block copolymer, *Science* **268**(5218), 1728–1731.

65. Dong, H., Lund, R. and Xu, T. (2015). Micelle stabilization via entropic repulsion: Balance of force directionality and geometric packing of subunit, *Biomacromolecules* **16**(3), 743–747.

66. Paramonov, S. E., Jun, H.-W. and Hartgerink, J. D. (2006). Self-assembly of peptide-amphiphile nanofibers: The roles of hydrogen bonding and amphiphilic packing, *J. Am. Chem. Soc.* **128**(22), 7291–7298.

67. Hartgerink, J. D., Beniash, E. and Stupp, S. I. (2001). Self-assembly and mineralization of peptide-amphiphile nanofibers, *Science* **294**, 1684–1688.

68. Li, Y., Wang, F. and Cui, H. (2016). Peptide-based supramolecular hydrogels for delivery of biologics, *Bioengineering and Translational Medicine* **1**, 306–322.

69. Wang, Y.-C., Li, Y., Sun, T.-M., Xiong, M.-H., Wu, J., Yang, Y.-Y. and Wang, J. (2010). Core–shell–corona micelle stabilized by reversible cross-linkage for intracellular drug delivery, *Macromol. Rapid Commun.* **31**(13), 1201–1206.

70. Nichols, J. W. and Bae, Y. H. (2014). EPR: Evidence and fallacy, *J. Control. Release* **190**, 451–464.

71. Chauhan, V. P. and Jain, R. K. (2013). Strategies for advancing cancer nanomedicine, *Nat. Mater.* **12**(11), 958–962.

72. Dobrovolskaia, M. A. and McNeil, S. E. (2007). Immunological properties of engineered nanomaterials, *Nat. Nanotechnol.* **2**(8), 469–478.

73. Nijhara, R. and Balakrishnan, K. (2006). Bringing nanomedicines to market: Regulatory challenges, opportunities, and uncertainties, *Nanomedicine: NBM* **2**(2), 127–136.

74. Mobley, A., Linder, S. K., Braeuer, R., Ellis, L. M. and Zwelling, L. (2013). A survey on data reproducibility in cancer research provides insights into our limited ability to translate findings from the laboratory to the clinic, *PLoS One* **8**(5), e63221.

CHAPTER 2

Bioactive Supramolecular Assembly of Peptide Amphiphiles to Control Cellular Fate

M. T. JEENA, HUYEON CHOI and JA-HYOUNG RYU*

Department of Chemistry,
Ulsan National Institute of Science and Technology (UNIST),
Ulsan 44919, Republic of Korea

2.1. Introduction

Molecular self-assembly is a ubiquitous process in which individual molecules spontaneously organize themselves into ordered aggregates under inter/intramolecular forces such as electrostatic interactions, hydrophobic interaction, aromatic stacking and hydrogen bonding.[1–3] The comprehensive demonstration of self-assembly process is made by the nature which comprises a vast variety of self-assembled nanomaterials (Fig. 2.1).[4] The self-assembly generates much of the functionalities of life, thus it could be regarded as fundamental process of nature. The DNA, basis of all living beings, exists as double helix in which two antiparallel strands wraps around one

---

*Corresponding author: jhryu@unist.ac.fr

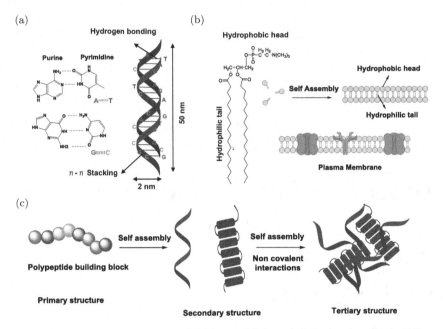

Fig. 2.1. Examples of biological self-assembled structure showing the building blocks and relevant interactions involved in the self-assembly process. (a) Double stranded DNA formed by the hydrogen bonding between the purine and pyrimidine base pairs. The π-π stacking also contributes to the formation of double helix structure. (b) Plasma-membrane of the eukaryotic cells, which is a phospholipid bilayer whose formation is driven by the hydrophobic-hydrophilic interactions. (c) The protein confirmation, where the higher order structures were driven by different non covalent interactions.

another via the strong hydrogen bonding of the base pairs (adenine-thymine and cytosine-guanine) which is further reinforced by the π-π stacking interactions between cyclic purines and pyrimidines.[5] The phospholipid bilayer of cell membranes is an inherent example for amphiphilic assembly where the hydrophobic tail tends to repel from the aqueous environment and hydrophilic heads point towards to form the cell membrane which isolate and protect the cell from the external environment.[6] The RNA and proteins fold to form the ribosomes which are the translationary machinery of the cell.[7] The cytoskeleton network of the cell constituting actin filaments, intermediate filaments and microtubules which are well organized assembly for regulating their functions.[8] The viruses which are proteins and nucleic acid based supra-molecules manifest their function through their self-assembly, as they form different size and shape such as rod, capsomers, and icosahedral.[7]

Tremendous efforts have been made till now to create a variety of synthetic functional nano-assemblies with "bottom-up" approach by following the nature's principles for organization of individual molecules. Different building units (biological and synthetic) are employed for creating interesting nanostructures which includes polymers, surfactants, nucleic acids, lipids, fatty acids and peptides. The peptides are promising materials for the biological application due to their existence in the biological system, biocompatibility and biodegradability.[9–11] Peptides are short oligomers made from 20 different amino acids with 20 different side chains which give their property in terms of charge, hydrophobicity, and functionality enables them to assemble into interesting structures with well-defined morphology in aqueous environment to form desired property.[12–14] In this chapter, we describe the basic principle underlies in the formation of peptide based nanostructure (Section 2.2) and their application in biomedical filed (Section 2.3). Even though the peptide nanostructure in bulk solvent and their applications are well explored over several decades, the *in situ* nanostructure formation inside a living cell is a pioneering strategy and found to have great potential in controlling the cellular fate. In Section 2.4, we describe the efforts on exploring the applications of intra cellular assembly and their impact on controlling the cell fate.

2.2. Self-assembly of Peptide Amphiphiles

Amphiphilicity (i.e. balance between hydrophilic and hydrophobic block) contributes a major role for the assembly behavior of peptides. The amphiphilicity triggers the formation of a variety of nanostructure ranging from micellar aggregates to bilayer structures in aqueous solution as they tend to minimize unfavorable interactions with the aqueous environment via aggregation process in which the hydrophilic domains become exposed whereas the hydrophobic domains are shielded.[15] The nanostructure formation (e.g. micelle, fiber, tube, dumbbell, rod, sheet) occurs via attractive forces such as hydrophobic interactions, hydrogen bonding, aromatic interaction or repulsive forces such as mechanical forces or electrostatic repulsions which give the secondary confirmations (alfa helix, parallel beta or antiparallel beta, random) to the Nano assemblies. There are two categories of peptide amphiphililes (PA); the first type includes peptide amphiphililes which comprises only natural amino acids of a long stretch of hydrophobic residue coupled to a number of polar amino acids or peptide with alternating hydrophobic and hydrophilic reside. These categories include peptides of the type $X_n Y_m$, in which **X** stands for hydrophobic units which serves as the tail group and **Y** stands for hydrophilic unit which act as the head. A Systematic study on the variation on the peptide of the form $X_6 K_n$ (X = alanine, valine or leucine; K = lysine, n = 1–5) showed different nanostructures vesicles to tube and ribbon.[16] The peptides $A_6 K$, $A_6 K_2$, $A_6 K_3$, $V_6 K_2$, $V_6 K_3$, $V_6 K_4$, $L_6 K_3$, $L_6 K_4$, and $L_6 K_5$ with amidated C terminus and acetylated N terminus exhibited similar confirmation in the CD spectra which is neither alfa helix or beta sheet, however formed different morphology depending on their hydrophobicity. The peptides $A_6 K$, $V_6 K_2$, and $L_6 K_3$ which are hydrophobic in nature formed nano tubes; $A_6 K_2$, $V_6 K_3$, and $L_6 K_4$ tend to form vesicles as they possess higher number of hydrophilic segments and $A_6 K_3$, $V_6 K_4$, and $L_6 K_5$ formed irregular aggregates (Fig. 2.2). The self-assembly behavior was controlled by the combined effects of hydrophobic and electrostatic interactions as well as hindrance effect established by the peptide sequence. These types of peptide nanostructures are promising carriers for the delivery of hydrophobic materials. An early studied example of these category is **Leu$_{12}$-Glu$_3$, Leu$_8$-Lys$_5$-Tyr and Fmoc-Leu$_8$-Lys$_5$-Tyr,** which forms alpha helix folds once spread in air-water interface and upon compression the molecules 'stood up', by lifting the apolar oligo leucine tail out of the water layer, and transformed into extended β-sheets.[17] The alternative placement of hydrophobic and hydrophilic unit also introduces amphihpilicity in the peptides which is also capable of forming verity of nanostructures. In the case of **Pro-Glu-(Phe-Glu)n-Pro** (n = 4, 5 or 7), the alternating hydrophobic phenylalanines and the hydrophilic glutamic acid residues caused the peptide chains to orient parallel to the water surface and form betaz confirmation.[18]

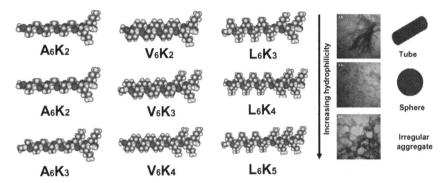

Fig. 2.2. Molecular structures of individual amphiphilic peptides. Color code: carbon, blue; hydrogen, gray; oxygen, red; nitrogen, yellow. Lengths of the peptides in their extended conformations are 2.7 (A6K), 3.2 (A6K2, V6K2, L6K2), 3.6 (A6K3, V6K3, L6K3), 3.9 (V6K4, L6K4), and 4.3 nm (L6K5). The peptide forms different nanostructure in accordance with the change in hydrophobicity, the most hydrophobic peptide showed nanotube formation and exhibited irregular aggregate with the increase in the hydrophilicity. (Reprinted with permission, ACS, 2012.)

The second category of peptide amphiphile includes the modification of peptides with a block of different properties such as lipids, polymers, alkyl chain, or fatty acids. Due to the surfactant nature, they orient the more hydrophilic peptide segment into the periphery, and the hydrophobic segment to the core of the assembly. *Stupp et al.* made remarkable contribution to explore the applications of this category of peptide amphiphiles over several decades. One among the early introduced amphiphilic peptide system consists of the incorporation of 16 carbon alkyl chain to the N-termini of the ionic peptide.[19,20] The peptide consists of three regions; A long alkyl chain (Region 1) which give amphiphilic property, a beta sheet forming domain (Region 2), and a charged head group (Region 3). The presence of charged elements into the peptide block, not only enhances their solubility in aqueous solutions, but also suppress their self-assembly as a result of charge repulsion (Fig. 2.3(a–c)). The net charge also endows the PA responsive to pH variations and salt addition and drives their self-assembly into cylindrical nanofibers upon neutralization or charge screening. It is reported that at higher pH, the screening of the repulsive forces on charged residues immediately

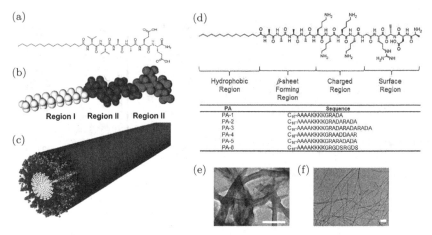

Fig. 2.3. Molecular structure of the investigated peptide amphiphile. (a) The molecule is composed of three segments, the alkyl tail, the β-sheet forming peptide sequence and the charged peptide head group. (b) Simulated structure for the peptide amphiphile showing three different regions. (c) Micellar structure induced by the peptide amphiphile. (d) Molecular structure for the peptide amphiphile possessing pH responsible sequence. (e)TEM images of the peptide amphiphile at pH 13 and 11. (Reprinted with permission).

adjacent to the beta sheet region is expected to allow the formation of parallel beta sheet that elongates the micelles into cylindrical nanofibers.

Tong *et al.* recently studied the lateral assembly of PA to form nanofiber to higher order structure by introducing pH responsive complimentary sequence in the PA.[21] In here, the 16 carbon containing alkyl chain is conjugated to amino acids sequence consisting of lysine, arginine and aspartic acid residue via alanine linker. The choice of lysine and arginine residues in the PA design allows selectively screen the charges of lysine residues at a pH between the pKa values of the two amino acids, thus isolating intrafiber interactions that elongate the nanostructure and interfiber interactions that promote lateral association of aligned nanofibers. The arginine and aspartic acid residues are separated by an alanine residue to prevent the neutralization of the side groups by proximity and thus allowing the interdigitation of surface region sequences when nanofibers align themselves in a supramolecular lateral self-assembly. Alanine was chosen over the conventional glycine as the spacer residue to reduce the flexibility and helical propensity of the surface region, thereby promoting interdigitation. Under the pH 4, the PA induces micelle morphology, and when the pH changed to 11 and 13, the nanostructure changed to fiber when the charged residue arranged in a complementary attractive fashion and fibrous bundles when the charged residue arranged in a complementary repulsive fashion (Figs. 2.3(d–f)). It is important to note that not all PA possess beta sheet confirmation, since additional factors can contribute in the determination of secondary structure. For example, the secondary structure of 19 different **C16G7ERGDS** PAs with *N*-methylated glycines at varying positions results in the observation when the modified amino acid was among the four closest to the core, fiber formation was blocked in the absence of hydrogen bonding between neighboring peptides. Further structural analysis revealed that while hydrogen bonding occurs near the core, it is possible for the amino acids in the periphery to adopt a weakly ordered conformation.[22,23]

2.2.1. *Bio catalytic peptide amphihpiles*

The bio catalytic or enzyme triggered self-assembly of peptides represents an alternative approach for the fabrication of adaptive nanostructures.[24] This approach allows detecting the presence of enzymes, screening enzyme inhibitors, assisting biomineralization, assaying the types of bacteria, and aiding the development of smart drug delivery systems. Xu *et al.* explored the enzymatic hydrogelation of small peptides and different application, where the enzyme involved includes phosphatase, thermolysin, β-lactamase, and phosphatase/kinase. Supramolecular hydrogels are inherently biodegradable and are being explored as alternative biomaterials, especially when the hydrogelators are bioactive molecules such as oligopeptides and their derivatives, for applications such as tissue engineering by using the hydrogels as the matrices of cell culture. Moreover, the self-assembled three dimensional fibril networks allow entrapment of both water molecules and other bioactive molecules (nutrients and proteins) or therapeutic agents. The enzymatic supramolecular

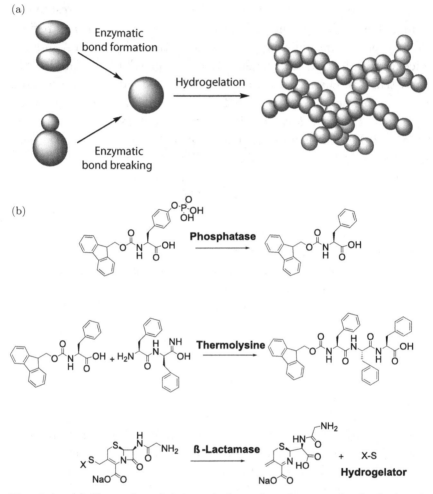

Fig. 2.4. (a) Illustration of enzymatic formation of supramolecular hydrogels via bond formation or bond cleavage. (b) Schematic representation of enzyme catalysed conversion of peptide amphiphile to self-assembling units.

hydrogelation involve both making and breaking bonds (Fig. 2.4(a)). Both routes allow the enzymes to convert the precursors to the hydrogelators, thus resulting in the self-assembly of hydrogelators in aqueous solution and the formation of supramolecular hydrogels. The phosphatases, an enzyme existing in organs, tissue and cells catalysis the removal of phosphate group from a substrate. Using this concept, the hydrogelator precursor can be designed with the equipment of phosphate group in the peptide building unit, whose assembly is inhibited due to the hydrophilicity. Upon, the enzymatic removal of phosphate group, the hydro gelation occurs with the formation of nanofibers with promising biomedical applications.[25]

For example, Fmoc-Tyr (phosphate) dissolves well in weak alkali aqueous solutions (a PBS buffer or a phosphate buffer). The addition of alkaline phosphatase to the solution converts **Fmoc-Tyr (phosphate)** to a more hydrophobic **Fmoc-Tyr**, which is a hydrogelator and results in a hydrogel. Xu *et al.* studied different variations in the peptide building unit which was equipped with tyrosine phosphate group to allow the hydrogelation upon enzymatic action. The phosphatases produce the hydrogel via bond breaking, however, the thermolysine involve in the bond making.[26] Thermolysin is a thermostable extracellular protease produced by bacteria, catalyzes the formation of a covalent bond via condensation or reversal of hydrolysis between two substrates (e.g., amino acids) in aqueous solution. The β-lactamases is a family of enzymes that hydrolyze the β-lactam antibiotics in bacterial strains. The hydrogelator design responsive for β-lactamases design involve incorporation of β-lactam ring in the precursor, since the β-lactamases catalyzes the opening of the β-lactam ring of the precursor, resulting in the subsequent rearrangement to release the hydrogelator, which self-assembles in water into nanofibers to afford a hydrogel (Fig. 2.4(b)).[27,28]

The bio catalytic hydrogelation is inspired from the dynamic process of living system as the biological components often interacts, assemble, compete and decompose when they are no longer required. The catalytic formation and degradation of actin filaments and microtubules are common example. As a mimic of such system, bio catalyst triggered peptide nanofibers with dynamic instability has been introduced by Uljin *et al.* recently, which offers a new opportunity for the creation of natural mimics.[29] For example, naphthoxy acetylene (Nap) modified dipeptide, **Nap-peptide-NH$_2$** which forms nanofiber and consequent hydrogel as a result of π-π stacking and hydrogen bonding has employed as the precursor in which the enzyme chymotrypsin catalyzes the forward reaction (under thermodynamic control) and the competing amide hydrolysis catalyzes the disassembly process resulting the dynamic instability

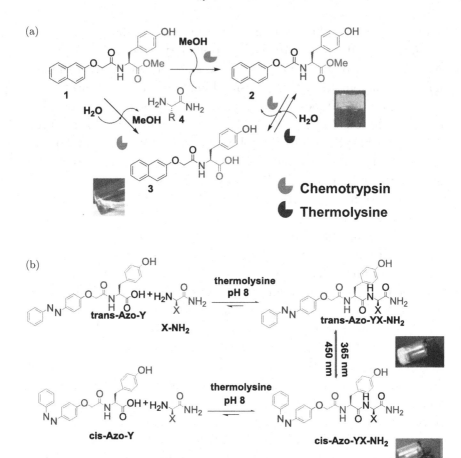

Fig. 2.5. (a) Nonequilibrium biocatalytic self-assembly. (a) Nap-Y-OMe (1) and different amino acid amides (X-NH₂, 4) in the presence of chymotrypsin (green) form the (temporary) hydrogelator (2) which may also be hydrolysed by the same enzyme to Nap-Y-OH (3) and (4). Thermolysin (blue) catalyzes the reversible hydrolysis/condensation of 2. (b) trans-Azobenzene substituted tyrosine derivatives (trans-Azo-Y) upon enzymatic condensation with amide derivatives of aminoacids (F-, L-, V-NH2) generate the corresponding dipeptide hydrogelators (Azo-YX-NH₂, X=F, CH₂–C₆H₅; L, CH₂–CH–(CH₃)₂; V, CH–(CH₃)₂) in presence of thermolysin at pH 8 in phosphate buffer under ambient conditions. (Reprinted with permission).

of the nanofiber (Fig. 2.5(a)). The requirement for creating such a system are: (i) the product of the forward reaction should have a tendency for unidirectional self-assembly, (ii) forward and backward reactions to produce/degrade gelator molecules should follow distinct routes, and (iii) conditions where the critical assembly concentration to form nanofibers is higher than the equilibrium yield of the enzymatic peptide synthesis reaction.[30] The advantage of bio catalytic assembly is the highly reproducible assemblies with optimized condition in such a way that, building unit of self-assembly sometimes cause poor water solubility resulting in the formation of kinetic aggregates whereas the biocatalysts mediated assembly gave thermodynamic products. The free energy change associated with molecular self-assembly and gelation is thermodynamically favorable to overcome the bias for peptide hydrolysis normally observed in aqueous systems to facilitate condensation which is proved with Fmoc and naphthoxy peptides. The incorporation Azobenzene moiety on the peptide building block showed the impact light responsiveness on the assembly. The equilibrium position of the amide reaction is controlled by the self-assembly propensity of the azo-peptide vs. its amino acid precursor. The azo-benzene is well known to form *cis-tras* isomers in responsive to the light. The trans azobenzene substituted tyrosine derivatives **(Azo-YX-NH$_2$)** generates hydrogelators upon enzymatic condensation in presence of thermolysin at pH 8 and the UV light exposure induce azo switching, resulting disassembly and hydrolysis of the gelator. The contributing interactions for the gelation are the π-π stacking between trans-azobenzene moieties, as well as the aromatic amino acids, combined with the hydrogen bonding interactions between the dipeptide units[31] (Fig. 2.5(b)).

2.2.2. *Nanostructure formation by following nature's principle*

The biological system is enriched with several folded protein systems which are essential for the existence of life; possess defined secondary confirmations in their structure which is also the basis for the creation of synthetic peptide nanostructures.[1] The primary structure of proteins constitutes amino acid sequence while the secondary structure possess different confirmations such as alpha helix, which the most common confirmation, first determined from the myoglobin proteins or beta strands formed via noncovalent forces (Fig. 2.6).[14] Mimicking the nature, the synthetic peptide assemblies also possess secondary confirmations and falls in either categories such as beta (parallel,

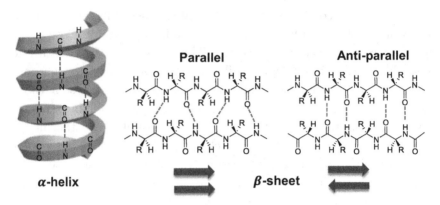

Parallel

Anti-parallel

α-helix

β-sheet

Fig. 2.6. Schematic representation showing alpha helix with the hydrogen-bonding pattern parallel to the helix axis and β sheet assembly is stabilized by the formation of intermolecular hydrogen bonds resulting in the formation both parallel and antiparallel confirmations.

anti-parallel, turns, hairpin or coil) or alpha (helices, coiled coil). The beta sheet assemblies are well known for their ability to assemble into fibrous structure observed in the several neurodegenerative diseases such as Parkinson's and Alzheimer's. The sequences from these fibril aggregates are adapted to create beta sheet forming peptide assemblies for developing new materials.[32]

One of the recent study on the aggregation process of amyloid protein suggest that the Aβ peptide of Alzheimer's diseases **Aβ(16–22) E22Q** or **Ac-^{16}KLVFFA^{22}Q-NH$_2$**, assembles as antiparallel β strands that later transition completely into parallel β strands.[33] The peptide **Aβ(16–22) E22Q** or **Ac-^{16}KLVFFA^{22}Q-NH$_2$** antiparallel strand orientation due to the secondary structure formation within the less hydrated peptide particle phase and then the electrostatic repulsion between the lysine side chains select against charged N-terminal lysine residue proximity in parallel strands (Fig. 2.7(a)). The phenyl alanine, the aromatic amino acid is important in triggering the self-assembly of amyloids due to their π-π stacking. They are also reported in the phenylketonuria as toxic fibril aggregates.[34] A model octapeptide consisting of alternating arginine and phenyl alanine **[Arg-Phe]$_4$** self-assemble into amyloid like beta sheet rich spheroidal aggregates as well fibrils[35,36] The peptide spontaneously form spheroid, at lower concentration

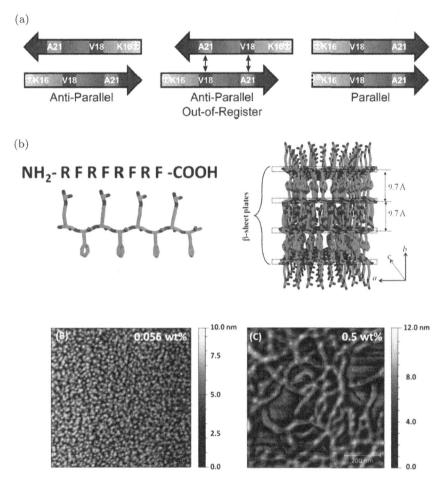

Fig. 2.7. (a) Strand conformations of Aβ (16−22) E22L peptide showing positions of charged lysine (blue) residues. Electrostatic repulsion is attenuated in antiparallel peptide orientation. Out-of-registry strands place the bulky valine packed with the less bulky alanine. Arrows indicate valine (red)-alanine (brown) cross-strand pairing. (b) Molecular model for [Arg-Phe]₄ (upper left), organization into β-sheets plates linked by π-stacking interactions and hydrogen bonding between side-chains (upper right), *ex situ* AFM imaging from [Arg-Phe]4 solutions at different concentrations, 0.056 wt.%, marginally above cac; and 0.5 wt.%, well above cac.

of 0.001 wt.% and transform into fibrils when the concentration increases to 0.17 wt.%. The peptide forms a cross-beta confirmation with anti-parallel strands separated by 4.8 A° perpendicularly arranged into beta sheets running parallel to the fiber long axis, in which the stacked beta sheets comprise phenyl alanine or arginine side chains which are kept together by strong π-π stacking interactions and hydrogen bonding (Fig. 2.7(b)). A precise control over the lamination and untwisting of one dimensional assemblies of beta sheet rich peptide of phenyl alanine are possible by tuning the terminal charges of the sequence. The charged hydrophilic amino acids such as lysine or glycine are frequently used for the modulation of electrostatic interactions in the peptide assemblies (Fig. 2.8).

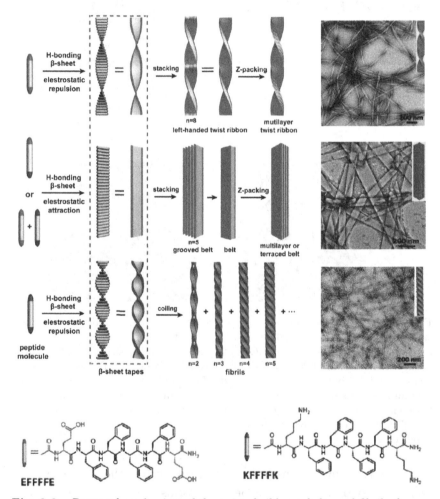

Fig. 2.8. Proposed mechanism of the twisted ribbons, belts and fibrils, formation by the peptides EF4E, EF4K, EF4E/KF4K mixture, and KF4K.

For examples the short designer peptides of **Phe** core which were equipped with charged terminus at both ends **EFFFFE** (both terminus contain negative charge) **KFFFFK** (both terminus contain positive charge) **EFFFFK** (positive and negative charge at C and N terminus respectively) showed perfect twisting in the beta sheet as well as morphologies depending on the charges.[36]

2.3. Post Assembly to Control Cellular Fate

2.3.1. *Bioactive assembled structure from peptide amphiphiles*

The modification of functional entity on the PA with biologically relevant epitopes could introduce bioactivity to the PA. For example, the introduction of **Arg-Gly-Asp (RGD)** is a well-known biological epitope presented in the cell binding domains of extracellular proteins such as fibrinectin and vitronectin, and is found to be bind with integrin receptors on the cell membrane. The modification of PA with **RGD** allows easy recognition of PA by the cell membrane.[37] A PA has designed with amino acid sequence **of KXXXAAAK** (X = G or L) which could form beta confirmation via hydrogen bonding resulting in the formation of long cylindrical micelle, the alkyl segment has introduced on the Lysine side chain at the C terminal for providing amphiphilicity. The **RGD** with biotin functionality has been introduced on the N terminal to allow the recognition by the cellular protein (avidin) for the improved bioactivity of PA.[38] The branched architecture of PA enhances the accessibility of epitope for protein binding and also allows the presentation of more than one epitope in a single molecule.

The epitope topography at the nanoscale structure of a scaffold influences its bioactive property independent on epitope density as well mechanical property. The change in the linker length between **RGDS** to the base peptide backbone results in the different bioactivity of the PA. The beta sheet forming peptide back bone **VAAAKKK**, which is functionalized with alkyl chain at the N terminus has been synthesized and conjugated with **RGD** at the C terminus via glycine linker in which the linker length changes as 1, 3 and 5.[39] The co assembly of PA with the peptide back bone in the ratio 1:9 by weight induces nanofiber with improved bioactivity since the co assembly reduces epitope crowding on the nanofiber surface. The variation of the linker length caused different spatial distribution of the epitope (Fig. 2.9(a)). The PA with long linker length cause enhanced fibroblasts response due to the efficient binding of the epitopes to the integrin receptors located on the cell membrane as a result of separation of epitope from the bulky charge backbone and spacing out the epitope from the surface of the nanofiber. However, the replacement of glycine with flexible PEG and rigid aromatic linker showed only minimal influence on the cell response, suggesting that the topography of the epitope could control the bioactivity. The

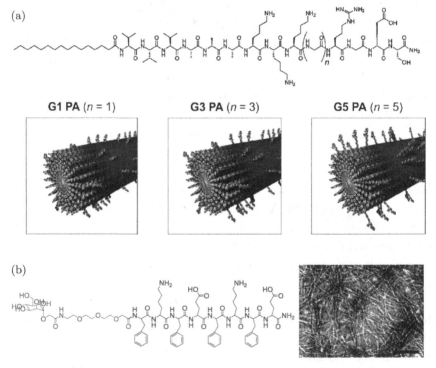

Fig. 2.9. (a) Chemical design of RGDS epitope-presenting PAs, where a glycine linker of variable length presents the epitope. Molecular graphics representation of RGDS epitope display on the co-assembled nano fiber surface, where the linker consists of one, three, or five glycine residues (named as G1 PA, G3 PA and G5 PA, respectively). (b) Structure of glycoconjugate peptide and corresponding TEM images.

greatest advantage of peptides which make them to be the interesting candidates for the biological response is the ability to modify the functional group. The replacement of RGD with the folic acid allows the PA to be responsive towards folate receptor. For example, the folic acid conjugation to the lysine side chain of **C16 V₃A₃K₃-NH₂** induces nanofiber formation with targeting ability to the cancer cell.

Apart from alkyl chain functionalized peptide amphiphile, the PEG functionalization of peptide assembling block is also found to enhance the self-assembling behavior.[40] For example, Lim *et al.* have synthesized the peptide amphiphilic consisting of **FKFEFKFE** as a beta sheet forming unit which is further conjugated with linear or dendritic mannose functionalized PEG, as a coil block (Fig. 2.9(b)). The linear PEG functionalization induced strong beta sheet nanoribbon with anti-parallel packing. However, the dendritic coil blocks interfere with then repetitive beta sheet growth of the peptide block because of steric crowding, thereby resulting in the premature termination of nanostructure growth. This nanoribbon is found to interact with E. coli (*E. coli* strain with mannose binding adhesion FimH in its type 1 pili (ORN 178)) and inhibits their growth.[41]

2.3.2. *Antibacterial activity of self-assembled peptides*

Antimicrobial peptides inhibit infectious microbes and multidrug-resistant bacteria and highly positively charged antimicrobial peptides showed broad spectra activities.[42] The formation of α-helical or β-sheet nanostructures can interact with negatively charged membranes or cell surfaces resulting in the disruption of cell membrane and promoting antimicrobial activity.[43–46] Liu *et al.* reported that self-assembled cationic nanoparticles can be a strong and efficient antimicrobial agent. The **TAT (YGRKKRRQRRR)** peptide which is cell-penetrating peptide required for membrane translocation with highly positive charges and a hydrophobic block of cholesterol (C) with a spacer of three glycine (G3), can drive the self-assembly and improve membrane permeability (Fig. 2.10(a)). These cationic nanoparticles (**CG3R6TAT**) possess antimicrobial activities which efficiently inhibit the growth of diverse gram-positive

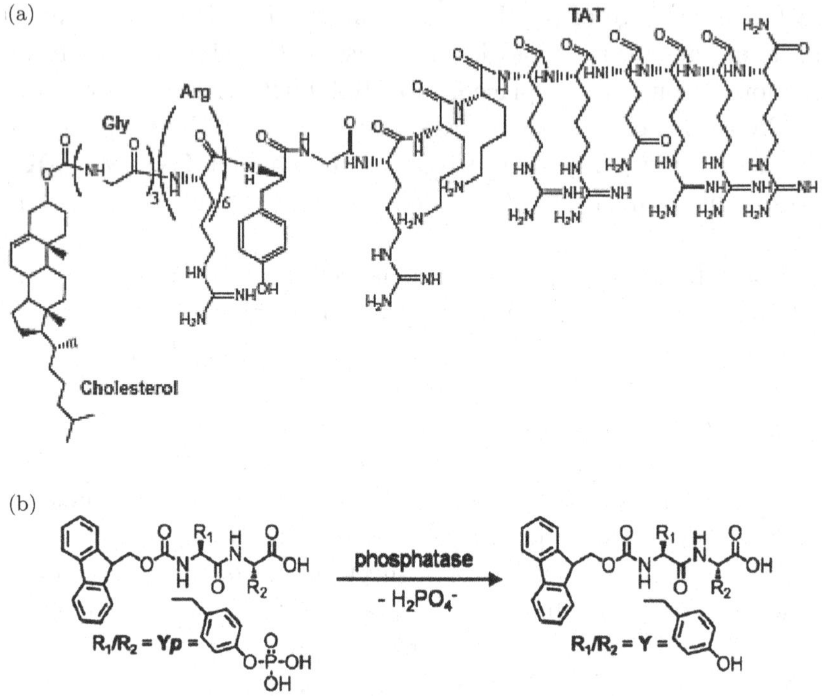

Compound	Amino Acid Sequence	Precursor Amino Acid Sequence
1	FY	FYp
2	YT	YpT
3	YS	YpS
4	YN	YpN
5	YQ	YpQ

Fig. 2.10. (a) Chemical structure of the antimicrobial peptide with cholesterol, three glycine and arginine based TAT (Self-assembled aromatic amphiphiles have antimicrobial properties responsive enzyme (b) Reaction scheme of dephosphorylation by alkaline phosphatase of Fmoc protected dipeptide amphiphiles.

even drug-resistant gram-positive bacteria. The nanoparticles have good therapeutic effect against *staphylococcus aureus* as a bacterial infection in brain over blood-brain barrier (BBB) limitation. The mechanism of efficient antimicrobial activity was confirmed by scanning electron microscopy (SEM) and transmission electron microscopy (TEM) images. The cationic nanoparticles were shown cellular uptake by means of non-specific electrostatic interactions which can make cell division and disrupt the cell wall, resulting in osmotic lysis of the cells. This peptide nanoparticle does not cause significant damage to the manor organs indicating that this nanoparticle is a highly efficient antimicrobial agent in infection regions.[42] In later, they showed that the nanoparticles have strong antimicrobial activity for treatment of *Cryptococcus neoformans* (yeast) induced brain infections over BBB.[47]

Laverty *et al.* described that an ultra-short aromatic-cationic peptides which can self-assemble nanofibers into hydrogel with antimicrobial activity. The five naphthalene based peptides (**NapKK, NapFFKK, NapFFK'K', NapFFOO** and **NapFF-FKK**) which formed the hydrogel showed increasing antibiofilm activity. The lysine containing peptides have significantly increasing selectivity against biofilm bacteria with 2% w/v **NapFFKK** hydrogels inhibiting the *Staphylococcys epidermidis* biofilm by 94%.[48] Similar to naphthalene, Hughes *et al.* reported that enzyme triggered self-assembly of Fmoc-protected short peptide can induce bacterial cell death by intracellular fibril formation.[49] Fmoc protected dipeptides including in **FY, YT, YS** and **YQ** containing phosphorylated precursors which are responsive to enzyme, alkaline phosphatase (Fig. 2.10(b)). Without the enzyme, the phosphorylated Fmoc-protected dipeptides cannot make any structures due to electrostatic repulsion. In presence of alkaline phosphatase, clear high concentration solution of phosphorylated peptides changes to foggy hydrogel composed of hydroxy-functionalized peptide via π-stacked Fmoc groups. The hydrophobic peptide (**Fmoc-FY-OH**) was found inside the cell due to accumulation by hydrophobic interaction with membranes. However, the three hydrophilic peptides (**Fmoc-YT-OH, Fmoc-YS-OH** and **Fmoc-YQ-OH**) are found in media. According to cell viability experiment, regardless of location and chemical structures of self-assembled nanofiber composed of aromatic peptides, the antimicrobial response has insignificantly different inside overexpressed enzyme *E. coli*.

2.3.3. *Drug delivery applications*

The conventional small drug molecules have the disadvantages due to poor solubility and low target selectivity. For anticancer drug delivery, the drug molecules were conjugated with peptides by biodegradable linker that can be cleaved by stimuli such as enzymes, pH and light. Peptide-drug amphiphiles have three essential parts including peptide, linker and drug. *Cui* and coworkers demonstrated supramolecular strategy that the anticancer drug camtothecin (CPT) conjugated with Tau-β-sheet forming peptide (Tau) via butyl disulfide (buSS) linker to form nanofibers (Fig. 2.11(a)). Based on different numbers of CPT in Tau peptide, the loading capacity was changed from 23% to 38%. These **mCPT-buSS-Tau**, **dCPT-buSS-Tau** and **qCPT-buSS-Tau** prodrugs assembled into nanostructures such as nanotubes. The disulfide linker of prodrug can be cleavage by glutathione (GSH) which is overexpressed inside cancer cells. Compared to only small CPT drugs, it has high selectivity to wider range of cancer cell lines with low IC_{50}.[50,51] In later, they have expanded this strategy to conjugate with different drugs like methotrexate (MTX), paclitaxel (PTX) and doxorubicin (DOX).[52–54] DOX conjugated with octa-arginine peptides which has high positively charged and BHQ-2 quencher via tetrapeptide **GLFG** linker which is cleavable under protease cathepsin B, overexpressed in many types of cancer. Before cleavage by cathepsin B, black hole quencher, BHQ-2, made quenching phenomenon of DOX fluorescence (Fig. 2.11(c–d)). In presence of cathepsin B, the tetrapeptide linker was cleavage to show 10 times higher emission of DOX that enzyme specific prodrug can take role as therapeutic agents.[55] Choi *et al.* synthesized simple short positive peptide (**Lys-Cys-Lys, KCK**) was conjugated with camptothecin (CPT) via disulfide linker as a prodrug (**KCK-CPT**). By forming complex with negatively charged biocompatible polymer, hyaluronic acid (HA) and positively charged **KCK-CPT** prodrug, the negative surface of the complex suppress the non-specific interaction by shielding charges to form the micelle like structure. Hyaluronidase (Hyal) and glutathione (GSH) which are overexpressed in cancer cells can degrade the outer shell of complex and disulfide bond to release free CPT. This complex (**HA-KCK-CPT**) internalizes only cancer cells not normal cells and induce apoptosis.[56]

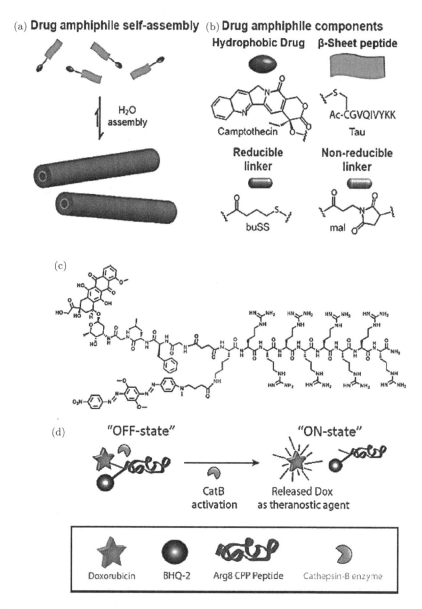

Fig. 2.11. (a) Schematic description of the designed and synthesized peptide-drug amphiphiles with self-assembled nanostructures. (b) Three component parts including in the hydrophobic drug CPT, Tau peptide and buSS cleavable linker (c–d) Enzyme-specific drug beacon containing Dox for drug-resistant theranostics (c) Chemical structure of the drug-beacon with octa-arginine sequence (purple), GFLG-linker (green), Dox (Red) and BHQ-2 (black) (d) Schematic illustration of drug beacon activation by Cathepsin B. (Reprinted with permission.)

2.3.4. *Biomacromolecule delivery of self-assembled peptides*

The self-assembled peptide nanostructures are used to encapsulate and deliver the large size of biomacromolecules such as proteins, antibodies, gene and polysaccharides. Among many candidates of self-assembled peptide nanostructures, hydrogels which are composed of hydrated nanofiber network are favorable to efficiently encapsulate and release the macromolecules by maintaining their bioactivity.[57] Fettis *et al.* reported that the synthetic peptide **Ac-QQKFQFQFEEQQ-Am** (Q11) which self-assembles to form nanofibers in aqueous solution.[58] However, it can be desolvated to form microgel under non-solvent such as ethanol. To encapsulate a model protein and deliver the protein drug delivery, they studied super-folder green fluorescent protein (**sfGFP**) and wheat germ agglutinin (WGA). Fluorescent sfGFP was released rapidly from Q11 microgel with burst kinetics. To possess micromolar affinity between WGA and Q11, Q11 was modified with glycosylation to make n-acetylglucosamine-Q11 (**GlcNAc-Q11**) which has high loading efficiency and controlled release of WGA proteins.

Branco *et al.* described that self-assembling peptides (MAX1 and MAX8) which make hydrogels made of β-hairpin under physiological pH and salt conditions.[57] In this work, they performed dextran encapsulation in β-hairpin hydrogels. The dextran molecules have different molecular weights and size such as 20 kDa (6.2 nm), 70kDa (9.1 nm) and 150 kDa (12.1 nm) respectively. Also, they studied that how macromolecular charge affects release the charged proteins, lactoferrin, which a 77 kDa protein and 7.2 nm hydrodynamic diameter. The smallest dextran (20 kDa) has fastest diffusion constant (D) in 0.5 wt.% as well as 2 wt.% MAX1 hydrogels compared to 150 kDa dextran, the slowest release. In electrostatic interaction view, negatively charged dextran and positively charged lactoferrin which are similar molecular weight and size have different release rate due to highly positively charged peptide networks. The negatively charged dextran has slower release rate than positively charged lactoferrin such 40% and 70%, respectively after 2 days.

Koutsopoulos *et al.* reported that **Ac-(RADA)$_4$-CONH$_2$** peptide hydrogel can be an efficient slow-release carrier of proteins

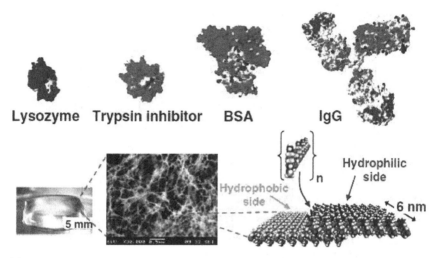

Lysozyme Trypsin inhibitor BSA IgG

Fig. 2.12. Representative description of functional proteins encapsulation in Ac-(RADA)$_4$-CONH2 peptide hydrogels. Proteins are lysozyme, trypsin inhibitor, BSA and IgG. The picture and its SEM image show fibrous network composed of self-assembled hydrophilic side and hydrophobic side (red; negatively charged, blue; positively charged, light blue; hydrophobic). (Reprinted with permission.)

including lysozyme, trypsin inhibitor, BSA and IgG (Fig. 2.12).[59,60] The four proteins have different molecular weight, hydrodynamic radius and isoelectric points (pI). Increasing molecular weight of proteins makes slow-release and small diffusion constants in PBS and gel. In addition, the self-assembled peptide nanostructures can deliver the immunomodulatory materials to desired region. The peptide-based vaccine carrier should protect the materials from degradation and stimulate the production of cytokines which make an immune response.[61] *Tirrell* and coworkers reported that the cylindrical micelles composed of lipopeptides (diC$_{16}$-OVA) take role as self-adjuvanting vaccine.[62] The **OVA** peptide is **EQLESIIN-FEKLTE** which is derived from cytotoxic T cell (T$_c$-cells) epitope, the model tumor antigen ovalbumin. The cylindrical micelles made of diC$_{16}$-OVA have 8.0 ± 2.3 nm in diameter and the length ranged from 50–300 nm. Without any additional adjuvants, this cylindrical micelle can stimulate the T$_c$-cells by anchoring the membranes with hydrophobic tails not activating the TLR2 proteins.

Peptides based self-assembled nanostructures have emerged as nanomedicine for delivering gene, DNA and RNA. Especially, positively charged amino acids interact with negatively charged oligo-nucleotides via electrostatic interactions, hydrogen-bonding and hydrophobic interactions. Arginine/histidine-rich peptide sequence can improve efficiency of gene delivery.[63] By forming nanocarriers such as micelles and hydrogels, the nucleic acids can be encapsulated inside the nanostructures and delivered to target site. Seow *et al.* reported cationic triblock amphiphilic peptides performing more efficient gene delivery carrier than polyethylenimine (PEI)[64] which is highly positively charged polymer commonly used for gene delivery. The three blocks consist of octaarginine blocks for binding DNA, four histidines to promote endosomal escape and hydrophobic amino acids such as F (**Phe**), W (**Trp**) and I (**Ile**). The size of three peptides is bimodal distribution with small particles (50–300 nm) and large particles (2–9 μm) and same as after binding with DNA (a luciferase reporter gene). These oligonucleotides show more efficient gene transfection and higher luciferase expression than PEI. Based on degree of hydrophobic amino acids, the luciferase expression level is different. In case of F (**Phe**), the luciferase expression level showed the most efficient in HEK 293 and HepG2 cells due to the highest hydrophobicity. In addition, in mouse breast-cancer model, I (Ile) oligopeptide shows 13 times higher luciferase expression level than that mediated by PEI. W (Trp) oligopeptides, lowest hydrophobic degree, produced the least expression level. Thus, the development of gene delivery can be further achieved by changing the length of amino acids or blocks.

Sigg *et al.* demonstrated stimuli-responsive co-delivery of nucleic acids and small drug molecules inside nanocarrier of the self-assembling peptides (H3SSgT) composed of three components: three histidines, reducible disulfide linker and hydrophobic L-tryptophan-D-leucine repeating unit derived from gramicidin A (gT) (Fig. 2.13).[65] This micelle nanoparticle can encapsulate BODIPY as a model of a hydrophilic drug and nucleic acids (antisense 22-mer oligonucleotides (**AON**) (**5'TAACAGGATTAGCA-GAGCGAGG3'**). The nanocarrier can be degraded by glutathione (GSH) inside cancer cells to release both cargos.

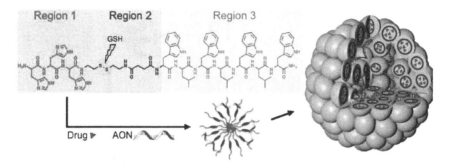

Fig. 2.13. Peptide design containing three parts for efficient gene delivery (a) Illustration of designed peptide amphiphile is composed of three components: Region 1. Hydrophilic part as three histidines (H), Region 2. Reducible disulfide linker, Region 3. Hydrophobic part including Trp (W) and Leu (L). (Reprinted with permission.)

2.3.5. *Tissue engineering*

Hydrogels were used as best biomaterials composed of fibers network system to provide a supportive 3D environment for cells to migrate, proliferate and adhere. The great numbers of articles related to tissue engineering with hydrogels have been reported for a few decades. Compared to polymer or other materials, peptides have biological benefits including in stimulating cell adherence or cell growth factor due to protein mimics.[66] Especially, the β-structured peptide to form fibrous network for hydrogel formation are mostly used for tissue-engineering by mimicking the extracellular matrix (ECM). For applying tissue applications, self-assembled peptides nanostructures should have the porous form for cell adhesion or growth. The short aromatic peptides self-assemble into fibrous structures by hydrogen bonding interactions and $\pi - \pi$ interaction in water. Wang *et al.* proposed that halogenated Fmoc-short peptides to form hydrogel can proliferate cells (Fig. 2.14).[67] The best candidate as a hydrogelator among **Fmoc-Phe** modified with fluoride, chloride and iodide is **Fmoc-4-fluoro-Phe** (Fmoc-fF) which can make gel at 0.15 wt.% concentrations. For enhancing cell adhesion and division, RGD peptide which can interact with integrin on cell surface is introduced to Fmoc-fFfF. This **FmocfFfFGRGD** can promote the proliferation of mouse embryonic fibroblast cells (NIH-3T3) with low cytotoxicity

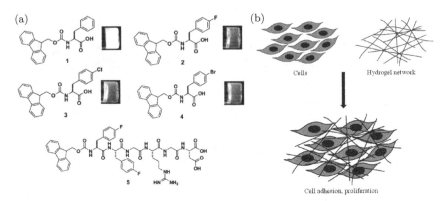

Fig. 2.14. Halogenated Fmoc-short peptides for tissue engineering (a) chemical structures and hydrogelation pictures of Fmoc-short peptides (1-5). (b) Schematic illustration of cell proliferation for tissue enignerring by self-assembled peptide hydrogel.

compared to other halogenated short peptides. *Uljin* and colleagues developed that three diphenylpeptides (**Boc-FF-COOH, Fmoc-FF-COOH,** and **Nap-FF-COOH**) form fibrous network via interlocked β-sheets and π-stacks. These fibrous networks have similar dimensions with ECM for l the chondrocyte cell culture.[68] In terms of stability, *Motamed et al.* introduced that β-peptide hydrogel which has high stability compared to α-amino acids and retains their bioactivity. β-peptide hydrogel showed suitable stiffness and environment for cells to adhere and proliferate. They developed the C14-acylated 14-helical structure of N-acetyl-β^3 peptides to form nanofibers for cell proliferation of neural cell line, SN4741.[69]

In nature, amino acids can exist as D- and L-forms but biomacromolecules such as proteins or ribosomes mostly composed of L-form amino acids. Thus, many researches of tissue engineering applications with the self-assembling peptides have used L-form amino acids due to biocompatibility.[70] However, in terms of stability, the L-form peptide bonds can be degraded easily by natural proteases but D-form peptide bonds cannot be degraded because enzymes cannot recognize D-peptides.[71] *Luo et al.* showed the first application of self-assembling peptide nanofibers composed of all D-form amino acids for three dimensional cell cultures.[72] They synthesized two peptides: one is **d-EAK16** peptide composed of all D-amino acids as

Ac-(AEAEAKAK)$_2$-CONH$_2$, the other is **l-EAK16** peptide composed of all L-amino acid as same sequence as d-EAK16. Both d- and l-EAK16 peptides make nanofiber network with different chirality. **d-EAK16** easily makes stronger nanofibers in presence of PBS. It indicates that a process of 3D cell culture can be handled quickly and this nanofiber can be affected easily by ionic strength. Importantly, there was no change in cell viability and apoptosis level in both d- and **l-EAK16** peptides. Thus, **d-EAK16** which has resistance to protease degradation can self-assemble into nanofiber scaffolds to support the cell growth. The **SMMC7721** cells can be entrapped in the nanofibers and grow well in real 3-D microenvironment.

Stupp and coworker studied the neuronal growth driven by peptide amphiphile (Fig. 2.15). In here the original peptide amphiphile

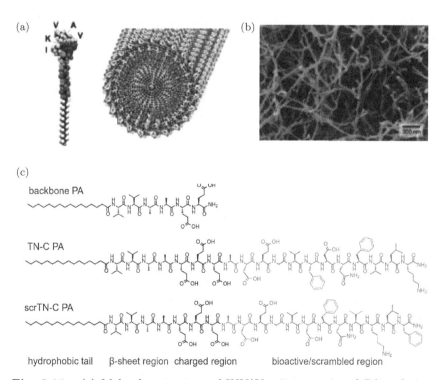

Fig. 2.15. (a) Molecular structure of IKVAV epitope equipped PA and simulated micelle morphology. (b) the TEM images (c) The chemical structure of TN-C-PA.

is modified with **IKVAV** epitope, a neurite promoting epitope, induced rapid differentiation of cells into neurons (murine neural progenitor cells (**NPCs**)). Further, the PA equipped with a peptide derived from extracellular glycoprotein tenascin-C mimetic (**TN-C-PA**) form supramolecular nanofibers. The **TN-C-PA** increased the migration of cells out of neurosphere cultured on gel coating. These bioactive gels could serve as artificial matrix in regions of neuronal loss to guide neural stem cells, promotes through biochemical cues neurite extension after differentiation.[13,73–75] *Xu et al.* introduced the self-assembly of glycol conjugate of peptide in the cell membrane surface as a mimic of glycoproteins which enhances the cell differentiation in murine embryonic stem cells. A self-assembling peptide consisting **Phe-Arg-Gly-Asp** has been synthesized which is conjugated with glucosamine and adenine, self-assemble into supramolecular assemblies that contain nanoparticle and nanofiber and doubled the cell proliferation of mES cells. The RGD in the design acts as integrin binding motif which is essential for the function of assembly. *Uljin et al.* recently introduced a dynamic surface modified with enzymatically active peptide group which showed enhanced adhesion for MSCs and proliferation. In here a plain glass surface where modified with salinization and PEG-ylation steps followed by solid-phase synthesis for attaching peptide with **RGD** motif and enzymatically cleavable site.[76–79]

2.4. *In-situ* Assembly to Control Cellular Fate

For the past decades, amphiphilic peptide based nanostructures have explored their applications in various area. However, in-situ self-assembly to control the cell fate have less investigated and quite recently attracted the attention. The observation that, several neurodegenerative diseases such as Parkinson, and Alzheimer's have associated with fibril aggregates of protein, researchers hypothesized that the peptide based fibrils could have a huge impact in controlling the cell fate either by their assembly near the cell or inside the cell which could be either beneficial or detrimental to the cell.[36] Moreover, certain endogenous protein monomers self-organize to certain structures such as microtubules, actin filaments, and vinculin which mediated indispensable cellular functions.

2.4.1. *Intracellular assembly of peptide amphiphile*

As a pioneer in the intracellular assembly, *Xu et al.* reported the enzyme triggered (alkaline phosphatase, ALP) fibrous assembly of small peptide based precursor amphiphiles, **NapFFK(NBD)Yp** inside of a living cell and successfully monitored the fibrous assembly with a fluorophore called 4-nitro-2,1,3-benzoxadiazole NBD, which exhibits bright yellow fluorescence upon the formation of fiber inside cell (Fig. 2.16).[80] The replacement of NBD with other fluorophore such as dansyl (DNS) or 4-(N,N-dimethylsulfamoyl)-2,1,3-benzoxadiazole (DBD) results in the different self-assembly behavior in vitro as well inside the cell, and thus this approach give

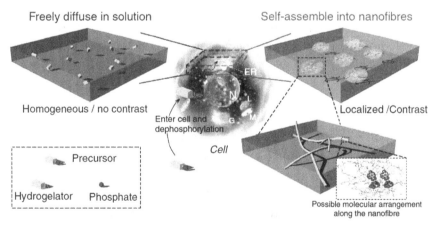

Fig. 2.16. The precursor turns to the corresponding hydrogelator after an enzymatic conversion inside the cell, a more hydrophobic molecule that self-assembles to form nanofibres at certain concentration. When the precursors are outside cells or the concentration of hydrogelator is too low to form nanofibres, those precursors or hydrogelators diffuse freely, distribute homogeneously, fluoresce identically within each pixel and thus show little contrast. Once the concentration of hydrogelator reaches high enough to form nanofibres, these nanofibres have more fluorophores within each pixel than the rest solution, and the fluorophores (as shown in CPK model) within nanofibres are localized, therefore the nanofibres fluoresce more brightly and generate the contrast. ER, endoplasmic reticulum; G, Golgi apparatus; L, lysosome; M, mitochondria; N, nucleus.

insight about the spatial distribution of nanofibers inside the cell and consequent impact on the cell function.[81] The nanofibers of dipeptides could contain phenyl alanine (Nap-Phe-Phe) could be selectively uptake by the cancer cells via micropinocytosis and interact with the cellular components such as cytoskeleton network and lead to the apoptosis of the cell by activating caspase.[82] *Gazit et al.* demonstrated that, aggregation of phenyl alanine itself at several millimolar concentration results in amyloid like nanofiber and induce toxicity which is found in the individuals with phenylketonuria. Diphenylalanine (FF) is a crucial building block in accelerating the amyloid-assembly process in Alzheimer and other 20 neurodegenerative diseases, which ultimately induces toxicity by the formation of toxic fibril aggregates. Thus, phenyl alanine containing peptide amphiphilic have been always gained therapeutic interest. Apart from peptide based fibrillary aggregate, a small molecule called **"1541"** which form nanofibrils showed partial localization in the lysosome of the living cell and induce partial leakage of the lysosome as well activate wide range of proteolysis and results in cell apoptosis by activating caspase as demonstrated by Well *et al.* which provide a general mechanism for fibril induced cellular toxicity.[83,84]

2.4.2. *Self-assembly in the pericellular space*

The formation of nanofiber on the cellular surface gains lots of interests. A tyrosine phosphate containing peptide amphiphile which is conjugated with naphthalene in the *N*-terminus, **Nap-Phe-Phe-Tyr(phosphate),** showed the effective inhibition of cancer cells with the formation of nanonets near in the pericellular space. **Nap-Phe-Phe-Tyr (phosphate),** which is precursor cleave effectively near the cell surface in the presence of ALP (dephosphorylation) and induce cell death upon hydrogelation with the formation of nanofiber. The pericellular formation of hydrogel also contributes to the entrapment of secretory phosphatase in the hydrogel which is main reason for the cell death. Certain carbohydrate amphiphiles which could form nanofiber is also found to inhibit the cancer cell progression. For example, glucose based amphiphiles found to form nanofiber near the cancer cell surface. The glucose based amphiphiles which equipped with a phosphate unit, *N*-**(fluorenylmethoxycarbonyl)-glucosamine-6-phosphate,** cleave in presence of ALP near ALP overexpressing cancer cell surface and transform the molecular structure to self-assembling monomeric unit *N*-**(fluorenylmethoxycarbonyl)-glucose** and effectively inhibits induce toxicity towards cancer cell upon nanofiber formation (Fig. 2.17).[85,86]

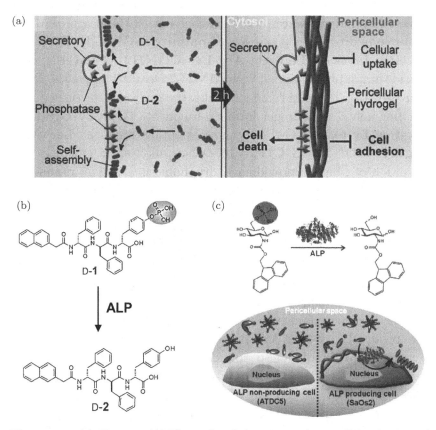

Fig. 2.17. (a) Enzyme (ALP) catalyzed formation of pericellular hydrogel/nanonets to induce cell death. (b) Molecular structures of the precursor (d-1) and the hydrogelator (d-2) (c) ALP catalyzed intracellular hydrogenation of carbohydrate amphiphile.

2.4.3. *Intracellular assembly triggered by enzyme*

Inducing fiber formation inside the cell required stimuli activation, of which enzyme activation is demonstrated until now, due to the plenty of available biological enzyme as well as availability of peptide sequence which could be activated or cleaved in presence of certain enzymes, which is an added advantage for the peptide based building unit.[87,88] The matrix metalloproteinase (MMP) is over expressed in the cancer cell compared with a normal cell. *Maruyama et al.* successfully demonstrated that the intracellular nanofiber formation with hydrogenation could successfully induce cancer specific toxicity with the usage of MMP cleavable hydrogelator precursor. **N-palmitoyl-Gly-Gly-Gly-His-Gly-Pro-Leu-Gly-Leu-Ala-Arg-Lys-CONH2 (ER-C16)** is a peptide amphiphile quipped with **MMP-7** cleavable sequence, **Pro-Leu-Gly-Leu**.[89] Upon cellular uptake, the hydrogelator turned into two fragments, a hydrogel forming fragment; **Gly-Gly-Gly-His-Gly-Pro-Leu-Gly (G-C16)** and a peptide fragment **Leu-Ala-Arg-Lys-CONH2**. G-C16, a 16 carbon alkyl chain enhances the formation of fiber by providing hydrophobic interaction in aqueous solution. The tetra peptide segment **Gly-Gly-Gly-His**, acts both acceptor as well donor of hydrogen bonding. However the fiber formation is prevented in the precursor **ER-C 16** due to the presence of cationic **Arg-Lys** unit which allow the molecules to behave as an efficient MMP transformable self-assembling unit and thereby induce cancer specific nano-fibrillation. The nanofiber induced significant toxicity towards cancer cell lines (Fig. 2.18).

The peptide conjugate of CBT (2-cyanobenzothiazole) derivative of taxol, **Ac-Arg-Val-Arg-Arg-Cys(StBu)-Lys(taxol)-2-cyanobenzothiazole,** also found to be effective against drug resistant cell lines upon their intracellular assembly to nanoparticles. The CBT is capable of undergoing a condensation reaction with GSH which is abundant inside the cancer cell. Liang *et al.* demonstrated the intracellular formation of self-assembling aggregates of CBT moiety. The rational design by the combination of biocompatible condensation and enzymatic self-assembly offered a new strategy for overcoming MDR.

The intracellular assembly of **Ac-Arg-Val-Arg-Arg-Cys(StBu)-Lys(taxol)-2-cyanobenzothiazole** into nanoparticles showed long lasting effect on tubulin condensation than simple taxol. Moreover, it showed 4.5 fold increments in anti-MDR effect on taxol-resistant HCT 116 cancer cell lines.[90] The intracellular fiber formation based on enzymatic assembly could also consider as an efficient tool for differentiating cancer cell environment. The combination of different enzyme activatable building unit in one molecular design allows this differentiation of cellular environment. For example, **Cys(SEt)-Glu-Tyr-(H₂PO₃)-Phe-Phe-Gly-CBT** consists of both ALP cleavable group, **Tyr-(H₂PO₃)** as well GSH reacting unit, CBT which allow the peptide amphihpile to sequentially respond to both ALP as well GSH and thus exhibits differentiable hierarchical self-assembly to nanofiber in the cellular environment. The enzyme ALP out of the cells or on the cell membrane catalytically dephosphorylates **Cys(SEt)-Glu-Tyr-(H₂PO₃)-Phe-Phe-Gly-CBT** to yield the hydrogelator **Cys(SEt)-Glu-Tyr-Phe-Phe-Gly-CBT** and further, once exocytosed by the cell, **Cys(SEt)-Glu-Tyr-Phe-Phe-Gly-CBT** undergo GSH induced condensation and form more stiff nanofiber with dimerformation which posess enhanced mechanical property. Since the GSH present inside the cell, the first formation of second fiber happens inside the cell (Fig. 2.19).[91]

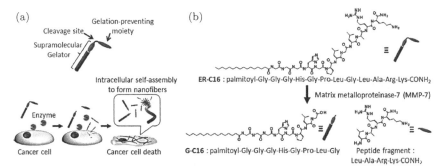

Fig. 2.18. (a) Molecular Structures of N-palmitoyl-Gly-Gly-Gly-His-Gly-Pro-Leu-Gly-Leu-Ala-Arg-Lys-CONH2(ER-C16), N-palmitoyl-Gly-Gly-Gly-His-Gly-Pro-Leu-Gly (G-C16), and Leu-Ala-Arg-Lys-CONH2 (Peptide Fragment). (Reprinted with permission.)

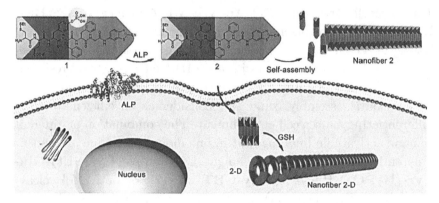

Fig. 2.19. Schematic illustration of ALP directed self-assembly of 1 into nanofiber 2 in extracellular environment and GSH-controlled condensation of 2 to yield the cyclic amphiphilic dimer 2-D which self-assembles into nanofiber 2-D in intracellular environment. Blue parts indicate the hydrophilic structures, and red parts indicate the hydrophobic structures. (Reprinted with permission.)

2.5. Summary and Future Outlook

The self-assembly provides an effective approach to construct discrete supramolecular nanostructures of various sizes and shapes in a simple manner. The peptide based nanostructures, constructed by following the principles of nature, have been widely employed over several decades in different biomedical field such as antimicrobial agent, tissue engineering, drug delivery etc. as the self-assembled peptide based nanostructure exhibits discrete property. Even though the applications of post assembled structures have studied widely, the *in situ* assembly of peptide amphiphile in the cell is quite recently introduced and found to be effective strategy to control cell fate. However, the achievement of spatiotemporal control over the self-assembly of molecules inside the cell is quite challenging due to the difficult to figure out the behavior of small molecules inside the cell due to its complexity. The self-assembly of peptide amphiphile inside the cell is often required activation by specific stimuli or enzyme, moreover the self-assembly required a certain concentration to which if often quite high which hamper the practical application of peptide based molecules. The localization of building block inside of certain predetermined cellular compartment could induce self-assembly without

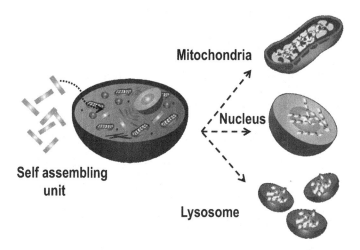

Fig. 2.20. Schematic representation for organelle localization induced self-assembly of peptide amphiphile.

any stimuli or harsh condition (Fig. 2.20). This approach of localization induced self-assembly provides novel outlook for disease therapy as well as in depth investigation of cellular functions.

References

1. Ulijn, R. V. and Smith, A. M. (2008). *Chem. Soc. Rev.* **37**, 664.
2. Lowik, D. W. and van Hest, J. C. (2004). *Chem. Soc. Rev.* **33**, 234.
3. Gazit, E. (2007). *Chem. Soc. Rev.* **36**, 1263.
4. Versluis, F., Marsden, H. R. and Kros, A. (2010). *Chem. Soc. Rev.* **39**, 3434.
5. Matsuurua, K. (2014). *RSC Adv.* **4**, 2942.
6. Petkau-Milroy, K. and Brunsveld, L. (2013). *Org. Biomol. Chem.* **11**, 219.
7. Mendes, A. C., Baran, E. T., Reis, R. L. and Azevedo, H. S. (2013). *Wiley Interdiscip. Rev. Nanomed. Nanobiotechnol.* **5**, 582.
8. Leckie, J., Hope, A., Hughes, M., Debnath, S., Fleming, S., Wark, A. W., Ulijn, R. V. and Haw, M. D. (2014). *ACS Nano* **8**, 9580.
9. Zhao, X., Pan, F., Xu, H., Yasseen, M., Shan, H., Hauser, C. A. E., Zhang, S. and Lu, J. R. (2010). *Chem. Soc. Rev.* **39**, 3480.
10. Cui, H., Webber, M. J. and Stupp, S. I. (2010). *Biopolymers* **94**, 1.
11. de la Rica, R. and Matsui, H. (2010). *Chem. Soc. Rev.* **39**, 3499.
12. Mandal, D., Nasrolahi Shirazi, A. and Parang, K. (2014). *Org. Biomol. Chem.* **12**, 3544.
13. Hamley, I. W. (2011). *Soft Matter* **7**, 4122.
14. Valéry, C., Artzner, F. and Paternostre, M. (2011). *Soft Matter* **7**, 9583.

15. Paramonov, S. E., Jun, H-W. and Hartgerink, J. D. (2006). *J. Am. Chem. Soc.* **128**, 7291.
16. Meng, Q., Kou, Y., Ma, X., Liang, Y., Guo, L., Ni, C. and Liu, K. (2012). *Langmuir* **28**, 5017.
17. Vankann, M., Möllerfeld, Jörg., Ringsdorf, H., Hartwig and Höcker (1996). *J. Colloid Interface Sci.* **178**, 241.
18. Rapaport, H., Möller, G., Knobler, C. M., Jensen, T. R., Kjaer, K., Leiserowitz, L. and Tirrell, D. A. (2002). *J. Am. Chem. Soc.* **124**, 9342.
19. Stendahl, J. C., Rao, M. S., Guler, M. O. and Stupp, S. I. (2006). *Adv. Funct. Mater.* **16**, 499.
20. Hartgerink, J. D., Beniash, E. and Stupp, S. I. (2001). *Science* **294**, 1684.
21. Chen, Y., Gan, H. X. and Tong, Y. W. (2015). *Macromolecules* **48**, 2647.
22. Trent, A., Marullo, R., Lin, B., Black, M. and Tirrell, M. (2011). *Soft Matter* **7**, 9572.
23. Paramonov, S. E., Jun, H-W. and Hartgerink, J. D. (2006). *J. Am. Chem. Soc.* **128**, 7291.
24. Yang, Z., Gu, H., Fu, D., Gao P., Lam, J. K. and Xu, B. (2004). *Adv. Mater.* **16**, 1440.
25. Yang, Z. and Xu, B. (2006). *Adv. Mater.* **18**, 3043.
26. Yang, Z., Liang, G. and Xu, B. (2007). *Soft Matter* **3**, 515.
27. Yang, Z., Ho, P-L., Liang G., Chow, K. H., Wang, Q., Cao, W., Guo, Z. and Xu, B. (2007). *J. Am. Chem. Soc.* **129**, 266.
28. Yang, Z., Liang G., Wang, L. and Xu, B. (2006). *J. Am. Chem. Soc.* **128**, 3038.
29. Debnath, S., Roy, S. and Ulijn, R. V. (2013). *J. Am. Chem. Soc.* **135**, 16789.
30. Abul-Haija, Y. M. and Ulijn, R. V. (2015). *Biomacromolecules* **16**, 3473.
31. Sahoo, J. K., Nalluri, S. K., Javid, N., Webb, H. and Ulijn, R. V. (2014). *Chem. Commun.* **50**, 5462.
32. Potapov, A., Yau, W. M., Ghirlando, R., Thurber, K. R. and Tycko, R. (2015). *J. Am. Chem. Soc.* **137**, 8294.
33. Liang, C., Ni, Rong., Smith, J. E., Childers, W. S., Mehta, A. K. and Lynn, D. G. (2014). *J. Am. Chem. Soc.* **136**, 15146.
34. Do, T. D., Kincannon, W. M. and Bowers, M. T. (2015). *J. Am. Chem. Soc.* **137**, 10080.
35. Decandio, C. C., Silva, E. R., Hamley, I. W., Castelletto, V., Liberato, M. S., Jr, O. and Alves, W. (2015). *Langmuir* **31**, 4513.
36. Hu, Y., Lin, R., Zhang, P., Fern, J., Cheetham, A. G., Patel, K., Schulman, R., Kan, C. and Cui, H. (2016). *ACS Nano* **10**, 880.
37. Guler, M. O., Hsu, L., Soukasene, S., Harrington, D. A., Hulvat, J. F. and Stupp, S. I. (2006). *Biomacromolecules* **7**, 1855.
38. Guler, M. O., Soukasene, S., Hulvat, J. F. and Stupp, S. I. (2005). *Nano Letters* **5**, 249.
39. Sur, S., Tantakitti, F., Matson, J. B. and Stupp, S. I. (2015). *Biomater. Sci.* **3**, 520.
40. Boekhoven, J., Zha, R. H., Tantakitti, F., Zhuang, E., Zandi, R., Newcomb, C. J. and Stupp, S. I. (2015). *RSC Advances* **5**, 8753.

41. Lim, Y. B., Park, S., Jeong, H., Ryu, J. H., Lee, M. S. and Lee, M. (2007). *Biomacromolecules* **8**, 1404.
42. Liu, L., Xu, K., Wang, H., K, P. J. T., Venkartraman, S. S., Li, L. and Yang, Y. Y. (2009). *Nature Nanotechnology* **4**, 457.
43. Zhou, J., Du, X., Li, J., Yamagata, N. and Xu, B. (2015). *J. Am. Chem. Soc.* **137**, 10040.
44. Bains, G. K., Kim, S. H., Sorin, E. J. and Narayanaswami, V. (2012). *Biochemistry* **51**, 6207.
45. Bains, G., Patel, A. B. and Narayanaswami, V. (2011). Pyrene: A probe to study protein conformation and conformational changes, *Molecules* **16**, 7909.
46. Sagara, Y. and Kato, T. (2009). *Nat. Chem.* **1**, 605.
47. Wang, H., Xu, K., Liu, L., Tan, J. P. K., Chen, Y., Li, Y., Fan, W., Wei, Z., Sheng, J., Yang, Y. Y. and Li, L. (2010). *Biomaterials* **31**, 2874.
48. Laverty, G., mcCloskey, A. P., Gilmore, B. F., Jones, D. S., Zhou, J. and Xu, B. (2014). *Biomacromolecules* **15**, 3429.
49. Hughes, M., Debnath, S., Knapp, C. W. and Ulijn, R. V. (2013). *Biomaterials Science* **1**, 1138.
50. Cheetham, A. G., Zhang, P., Lin, Y. A., Lock, L. L. and Cui, H. (2013). *J. Am. Chem. Soc.* **135**, 2907.
51. Cheetham, A. G., Ou, Y. C., Zhang, P. and Cui, H. (2014). *Chemical Communications* **50**, 6039.
52. Perry, S. W., Norman, J. P., Barbieri, J., Brown, E. B. and Gelbard, H. A. (2011). Mitochondrial membrane potential probes and the proton gradient: A practical usage guide, *Biotechniques* **50**, 98–115, doi:10.2144/000113610.
53. Cheetham, A. G., Zhang, P., Lin, Y. A., Lin, R. and Cui, H. (2014). *Journal of Materials Chemistry B* **2**, 7316.
54. Weinberg, S. E. and Chandel, N. S. (2015). Targeting mitochondria metabolism for cancer therapy, *Nat. Chem. Biol.* **11**, 9–15, doi:10.1038/nchembio.1712.
55. Lock, L. L., Tang, Z., Keith, D., Reyes, C. and Cui, H. (2015). *ACS Macro Letters* **4**, 552.
56. Choi, H., Jeena, M. T., Palanikumar, L., Jeoung, Y., Park, S., Lee, E. and Ryu, J. H. (2016). *Chem. Commun.* **52**, 5637.
57. Branco, M. C., Pochan, D. J., Wagner, N. J. and Schneider, J. P. (2009). *Biomaterials* **30**, 1339.
58. Fettis, M. M., Wei, Y., Restuccia, A., kurian, K. J., Wallet, S. M. and Hundalla, G. A. (2016). *J. Mater. Chem. B* **4**, 3054.
59. Koutsopoulos, S., Unsworth, L. D., Nagai, Y. and Zhang, S. (2009). *Proceedings of the National Academy of Sciences of the United States of America* **106**, 4623.
60. Koutsopoulos, S. and Zhang, S. (2012). *Journal of Controlled Release: Official Journal of the Controlled Release Society* **160**, 451.
61. Reed, S. G., Bertholet, S., Coler, R. N. and Friede, M. (2009). *Trends in Immunology* **30**, 23.
62. Black, M., Bertholet, S., Coler, R. N. and Friede, M. (2012). *Advanced Materials* **24**, 3845.

63. Hosseinkhani, H., Hong, P. D. and Yu, D. S. (2013). *Chemical Reviews* **113**, 4837.
64. Seow, W. Y. and Yang, Y.-Y. (2009). *Advanced Materials* **21**, 86.
65. Sigg, S. J., Postupalenko, V., Duskey, J. T., Palivan, C. G. and Meier, W. (2016). *Biomacromolecules* **17**, 935.
66. Rad-Malekshahi, M., Lempsink, L., Amidi, M., Hennink, W. E. and Mastrobattista, E. (2016). *Bioconjugate Chemistry* **27**, 3.
67. Wang, Y., Zhang, Z., Xu, L., Li, X. and Chen, H. (2013). *Colloids and Surfaces. B, Biointerfaces* **104**, 163.
68. Motamed, S., Del Borgo, M. P., Kulkarni, K., Habila, N., Zhou, K., Perlmutter, P., Forsythe, J. S. and Aguilar, M. I. (2016). *Soft Matter* **12**, 2243.
69. Berns, E. J., Sur, S., Pan, L., Goldberger, J. E., Suresh, S., Zhang, S., Kessler, J. A. and Stupp, S. I. (2014). *Biomaterials* **35**, 185.
70. Li, A., Hokugo, A., Yalom, A., Berns, E. J., Stephanopoulos, N., McClendon, M. T., Segovia, L. A., Spigelman, I. and Stupp, S. I. (2014). *Biomaterials* **35**, 8780.
71. Berns, E. J., Alvarez, Z., Goldberger, J. E., Boekhoven, J., Kessler, J. A., Kuhn, H. G., Stupp, S. I. *et al.* (2016). *Acta Biomaterialia* **37**, 50.
72. Newcomb, C. J., Sur, S., Lee, S. S., Yu, J. M., Zhou, Y., Snead, M. L. and Stupp, S. I. (2016). *Nano Letters* **16**, 3042.
73. Du, X., Zhou, J., Guvench, O., Sangiorgi, F. O., Li, X., Zhoi, N. and Xu, B. (2014). *Bioconjugate Chem.* **25**, 1031.
74. Roberts, J. N., Sahoo, J. K., McNamara, L. E., Burgess, K. V., Yang, J., Alakpa, E. V., Anderson, H. J., Hay, J., Turner, L-A., Yarwood, S. J., Zelzer, M., Oreffo, R. O. C., Ulijin, R. V. and Dalby, M. J. (2016). *ACS Nano* **10**, 6667.
75. Faruqui, N., Bella, A., Ravi, J., Ray, S., Lamarre, B. and Ryadnov, M. G. (2014). *J. Am. Chem. Soc.* **136**, 7889.
76. Gao, Y., Shi, J., Yuan, D. and Xu, B. (2012). *Nat. Commun.* **3**, 1033.
77. Gao, Y., Kuang, L., Du, X., Zhoi, J., Chandran, P., Horkay, F. and Xu, B. (2013). *Langmuir* **29**, 15191.
78. Roberts, J. N., Saho, J. K., McNamara, L. E., Burgess, K. V., Yang, J., Alakpa, E. V., Anderson, H. J., Hay, J., Turner, L. A., Yarwood, S. J., Zelzer, M., Oreffo, R. O. C., Uljin, R. V. and Dalby, M. J. (2016). *ACS Nano* **10**, 6667.
79. Faruqui, N., Bella, A., Ravi, J., Ray, S., Lamarre, B. and Ryadnov, M. G. (2014). *J. Am. Chem. Soc.* **136**, 7889.
80. Gao, Y., Shi, J., Yuan, D. and Xu, B. (2012). *Nat. Commun.* **3**, 1033.
81. Gao, Y., Kuang, Y., Du, X., Zhou, J., Chandran, P., Horkay, F. and Xu, B. (2013). *Langmuir* **29**, 15191.
82. Kuang, Y., Long, M. J., Zhou, J., Shi, J., Gao, Y., Xu, C., Hedstrom, L. and Xu, B. (2014). *J. Biol. Chem.* **289**, 29208.
83. Adler-Abramovich, L., Vaks, L., Carny, O., Trudler, D., Mango, A., Caflisch, A., Frenkel, D. and Gazit, E. (2012). *Nat. Chem. Biol.* **8**, 701.

84. Julien, O., Kampmann, M., Bassik, M. C., Zorn, J. A., Venditto, V. J., Shimbo, K., Agard, N. J., Shimada, K., Reingold, A. L., Stockwell, B. R., Weissman, J. S. and Wells, J. A. (2014). *Nat. Chem. Biol.* **10**, 969.
85. Pires, R. A., Abul-Haija, Y. M., Costa, D. S., Novoa-Carballal, R., Reis, R. L., Ulijin, R. V. and Pashkuleva, I. (2015). *J. Am. Chem. Soc.* **137**, 576.
86. Kuang, Y., Shi, J., Yuan, D., Alberti, K. A., Xu, Q. and Xu, B. (2014). *Angew. Chem. Int. Ed.* **53**, 8104. (82 same)
87. Li, J., Kuang, Y., Shi, J., Zhou, J., Medina, J. E., Zhou, R., Yuan, D., Yang, C., Wang, H., Yang, Z., Liu, J., Dinulescu, D. M. and Xu, B. (2015). *Angew. Chem. Int. Ed.* **54**, 13307.
88. Zhou, J., Du, X., Yamagata, N. and Xu, B. (2016). *J. Am. Chem. Soc.* **138**, 3813.
89. Tanaka, A., Fukuoka, Y., Morimoto, Y., Honjo, T., Koda, D., Goto, M. and Maruyama, T. (2015). *J. Am. Chem. Soc.* **137**, 770.
90. Yuan, Y., Wang, L., Du, W., Ding Z., Zhang, J., Han, T., An, L., Zhang, H. and Liang, G. (2015). *Angew. Chem. Int. Ed.* **54**, 9700.
91. Zheng, Z., Chen, P., Xie, Maolin., Wu, C., Luo, Y., Wang, W., Jiang, J. and Liang, G. (2016). *J. Am. Chem. Soc.* **138**, 11128.

CHAPTER 3

Supramolecular Gel Electrophoresis of Protein

MASAMICHI YAMANAKA*

Department of Chemistry, Shizuoka University, Japan

3.1. Introduction

Protein electrophoresis is one of the most frequently used techniques in life sciences research.[1] The technique has been dramatically evolved since the electrophoresis of serum albumin reported by Tiselius in 1937,[2] although several studies regarding protein electrophoresis have been reported in the nineteenth century. Protein electrophoresis was initially performed in free boundary conditions. Subsequent electrophoretic techniques were developed using paper or gel as a carrier, i.e., zone electrophoresis during the 1950s.[3] Polyacrylamide gel, which can be prepared by copolymerization of acrylamide and N,N'-methylenebisacrylamide (bis), was first applied as a carrier for protein electrophoresis in 1959.[4] Since then, protein electrophoresis using a polyacrylamide carrier has grown rapidly, and has become an indispensable technique for protein analysis for more than half a century. In particular, sodium dodecyl sulfate-polyacrylamide gel electrophoresis (SDS-PAGE), the procedure for which was first established in the early 1970's,[5] is the most frequently used method of

---
*Corresponding author: yamanaka.masamichi@shizuoka.ac.jp

protein electrophoresis. In SDS-PAGE, negatively charged denatured proteins are electrophoresed in the polyacrylamide gel, and separated mainly according to their molecular weights. Several useful derivative electrophoretic techniques such as affinity electrophoresis and two-dimensional (2D) electrophoresis have also been developed.[6,7] For instance, phosphorylated protein receptor (Phos-tag) copolymerized with polyacrylamide gel allows affinity electrophoresis of phosphorylated proteins to be performed.[8] 2D electrophoresis is a principal technique in proteomics research.[9] In spite of all the technical developments in protein electrophoresis, the development of novel carriers for protein electrophoresis is rarely reported. It is crucial that novel types of gel carrier for protein electrophoresis are created in order for the progression of electrophoretic techniques to continue. Therefore, we aimed to develop methods for protein electrophoresis using supramolecular gels, i.e., _Supramolecular Gel Electrophoresis_ (SUGE).

Small molecules called low molecular weight gelators form supramolecular gels through self-assembling noncovalent interactions.[10–14] The potential applications of supramolecular gels in the field of materials science have attracted researchers in recent decades.[15] In particular, supramolecular hydrogels, which immobilize an aqueous phase, have been developed as matrices of various biological applications such as enzymatic analysis, tissue engineering, and cell culture.[16–18] Although much attention has been paid to the biological applications of supramolecular hydrogels, no examples of protein electrophoresis using supramolecular hydrogels have been known until our first report in 2011.[19] Supramolecular hydrogels have potential advantages as carriers of electrophoresis in comparison with ordinary polyacrylamide gel. A serious drawback of PAGE is the difficulty in visualizing protein samples from gels after electrophoresis. Stimuli responsive gel to sol phase transition, a typical characteristic found in supramolecular gels, allows the efficient recovery of proteins. The structural diversity of low molecular weight gelators allows affinity electrophoresis to be tailored according to proteins of particular interest.

3.2. Supramolecular Gel Electrophoresis (SUGE) for Denatured Proteins

3.2.1. *Development of SDS-SUGE*[19]

We serendipitously discovered C_3-symmetric tris-urea molecule which functioned as a low molecular weight gelator for organic solvents.[20] The derivatives allow for various types of supramolecular gels.[21] An amphiphilic tris-urea **1** was designed as a low molecular weight gelator for aqueous medium (Fig. 3.1). Synthesis of **1** was achieved by a multistep reaction initiated by trimesic acid and α-D-glucose. Supramolecular hydrogel was formed by mixing **1** and Tris[tris(hydroxymethyl)aminoethane]-glycine-SDS (TGS) buffer (Tris: 25 mM; glycine: 192 mM; SDS: 3.5 mM), which is a typical buffer used in SDS-PAGE.

After a range of investigations, the following procedure was setup for SDS-SUGE to separate denatured protein samples (Fig. 3.2). 1) A supramolecular hydrogel was prepared from **1** (2.0 wt%), agarose (2.0 wt%), and TGS buffer. Agarose, whose gel itself has no separation ability for proteins in the range from 16–229 kDa, was added to

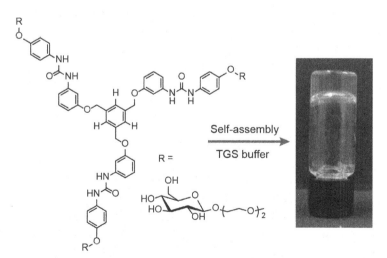

Fig. 3.1. Chemical structure of amphiphilic tris-urea **1** and photograph of the TGS buffer gel.

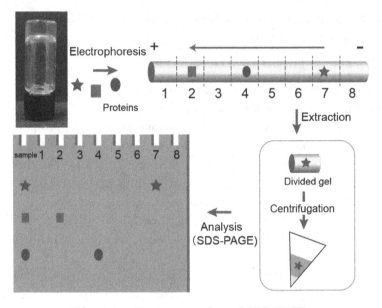

Fig. 3.2. Typical procedure of SDS-SUGE.

reinforce the physical strength of the supramolecular hydrogel. 2) A glass capillary (with ϕ 2 mm, length 120 mm) was filled up to 80 mm with the supramolecular hydrogel. 3) Sample protein solution was adsorbed onto one end of the hydrogel, and both ends of the capillary were filled with agarose gel (2.0 wt%). 4) The capillary was sunk in TGS solution in the submarine electrophoresis system and electrophoresed using optional voltage and time. 5) The electrophoresed gel was then removed from the glass capillary and divided into eight equal parts (numbered 1–8 from the anode), and proteins were isolated from the supramolecular hydrogel by centrifugation. As a remarkable characteristic, the supernatant resulting from simple centrifugation (14,100 g) of the electrophoresed supramolecular hydrogel contained about 50% of protein. 6) The separation pattern of the supernatants was analyzed using a standard SDS-PAGE system followed by Coomassie Brilliant Blue (CBB) staining. Six proteins, namely β-galactosidase (β-Gal, 116 kDa), bovine serum albumin (BSA, 66 kDa), ovalbumin (OVA, 45 kDa), carbonic anhydrase (CA, 29 kDa), lysozyme (LZM, 14 kDa), and aprotinin

(APR, 6.5 kDa) were used in SDS-SUGE tests. Typically, a mixture of two proteins with different molecular weights was applied in each experiment.

Several SDS-SUGE experiments revealed a unique separation manner of denatured proteins (Fig. 3.3). A mixture of β-Gal and OVA was employed for SDS-SUGE at 100 V for 120 min (Fig. 3.3a). Smaller OVA (45 kDa) was detected in lanes 4 to 7, and a stronger band was found in lane 5. Larger β-Gal (116 kDa) was detected in lanes 6 to 8, with the strongest band observed in lane 7. OVA was electrophoresed to a large extent on the anode side than β-Gal as

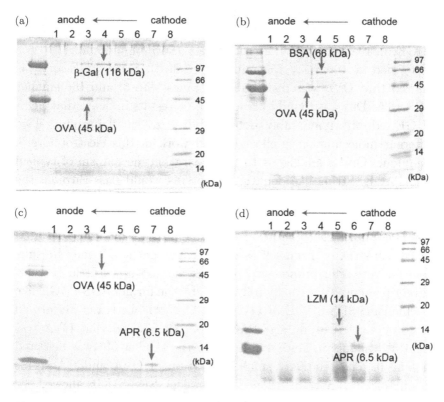

Fig. 3.3. SDS-PAGE analyses of SDS-SUGE separation using TGS buffer (25 mM Tris, 192 mM glycine, 3.5 mM SDS) of a) β-Gal (116 kDa) and OVA (45 kDa); b) BAS (66 kDa) and OVA (45 kDa); c) OVA (45 kDa) and APR (6.5 kDa); d) LZM (14 kDa) and APR (6.5 kDa).

well as standard SDS-PAGE; however, the separation was unclear. SDS-SUGE of BSA and OVA exhibited similar results (Fig. 3.3(b)). Smaller OVA (45 kDa) was detected in lanes 3 and 4, and a stronger band was found in lane 3. Larger BSA (66 kDa) was detected in lanes 4 and 5, with the strongest band observed in lane 4. Next, we chose a combination of OVA and the much smaller LZM. SDS-SUGE was performed at 100 V for 170 min. OVA (45 kDa) was detected in lanes 3 and 4, whereas the smaller LZM (14.4 kDa) was detected in lane 5. This result means that smaller LZM was retained closer to the cathode side compared with larger OVA, i.e., SDS-SUGE using the supramolecular hydrogel of 1 yielded a separation manner that was different from that observed in typical SDS-PAGE. To identify the nature of this unusual separation manner, APR was used in SDS-SUGE, instead of LZM. Electrophoresis of ovalbumin and APR was performed at 100 V for 150 min (Fig. 3.3(c)). SDS-PAGE analysis showed that OVA was distributed in lanes 3 to 5, and the smaller APR (6.5 kDa) was found in lane 7. These results indicate that SDS-SUGE exhibits remarkably poor mobility for small proteins. Two different molecular sieve effects would work in this electrophoretic technique. One is analogous to the separation mechanism of typical SDS-PAGE. String-like denatured proteins would pass through the three-dimensionally intertwining fibrous network of the supramolecular hydrogel, and smaller proteins would show large mobility. The other resembles the separation mechanism of gel filtration. Isolated spaces constructed from self-assembled aggregates of 1 may be suitable for retaining proper size proteins against an electric current. Larger proteins were chiefly influenced by the former SDS-PAGE like mechanism. SDS-SUGE of OVA and CA (29 kDa) was performed at 100 V for 170 min. SDS-PAGE analysis showed that OVA was identified mainly in lanes 4 and 5, and the smaller CA was detected in lane 6. The relatively small separation of OVA and CA seems to support the above-mentioned mechanism. SDS-SUGE allowed the separation of small-size proteins. A mixture of LZM and APR was applied for SDS-SUGE at 100 V for 180 min (Fig. 3.3(d)). SDS-PAGE analysis showed that LZM was mainly found in lane 5, and smaller APR was retained at the more cathodic lanes 6 and 7.

3.2.2. *Effect of ionic surfactant concentration on SDS-SUGE*[22,23]

An ionic surfactant such as SDS is essential to the formation of supramolecular hydrogel of **1**, and a mixture of **1** and pure water forms an insoluble suspension.[22] Various ionic surfactants, including not only anionic sodium sulfates but also cationic surfactants such as dodecyltrimethylammonium bromide, effectively induce hydrogel formation of **1**. However, non-ionic surfactants such as *n*-octyl-*β*-D-glucopyranoside and tetraethyleneglycol monododecyl ether could not induce gelation. The concentration of SDS was important for the formation of supramolecular hydrogel of **1**. The mixtures of **1** with low concentrations of SDS formed insoluble suspensions, and the mixtures of **1** with high concentrations of SDS formed clear solutions. SDS solutions within the range of 0.5–4 mM provided supramolecular hydrogels for 9 mM of **1**. The effective concentration of SDS was lower than the critical micelle concentration of SDS (8.2 mM). It was found that hydrogels that are more transparent were obtained using higher rather than lower concentrations of SDS. The differences in the transparency of the hydrogels of **1** are thought to reflect the rate of bundling. Scanning electron microscope (SEM) observations of mixtures of **1** and SDS showed fibrous aggregates of varying thickness, depending on the concentration of SDS (Fig. 3.4). The SEM image of a sample prepared from a suspension of **1** showed a shapeless morphology (Fig. 3.4a). The xerogel prepared from a hydrogel (9.0 mM of **1** and 0.50 mM of SDS) demonstrated intertwining fibrous aggregates ranging in thickness from 100–600 nm (Fig. 3.4b). The SEM image of the xerogel prepared from a hydrogel (9.0 mM of **1** and 4.0 mM of SDS) showed many homogeneous fibrous aggregates with diameters ranging from 100–250 nm (Fig. 3.4c). Notably, a dried sample of a solution (9.0 mM of **1** and 10 mM of SDS) showed fibrous aggregates with a large aspect ratio, and diameters of 50−80 nm (Fig. 3.4d). Similar results were observed for mixtures of **1** and the cationic surfactant (dodecyltrimethylammonium bromide). Amphiphilic **1** self-assembles and randomly bundles in aqueous media, but ionic surfactants appeared to restrict the rate of bundling.

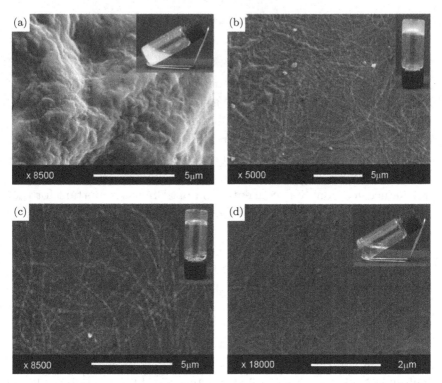

Fig. 3.4. Photographs and SEM images of a) insoluble suspension of **1** (9.0 mM); b) hydrogel of **1** (9.0 mM) and SDS (0.50 mM); c) hydrogel of **1** (9.0 mM) and SDS (4.0 mM); d) solution of **1** (9.0 mM) and SDS (10 mM).

The major separation manner of SDS-SUGE using TGS buffer (Tris: 25 mM; glycine: 192 mM; SDS: 3.5 mM) was altered by the molecular weights of proteins. For proteins with molecular weight >45 kDa, smaller proteins show better mobility than larger proteins, as in SDS-PAGE. In contrast, for proteins with molecular weight <45 kDa, smaller proteins show poorer mobility than larger proteins. Moreover, the thickness of the fibrous aggregates in the supramolecular hydrogel of **1** changes depending on the concentration of SDS. These observations led us to hypothesize that the mechanism of protein separation in SDS-SUGE depends on the concentration of SDS.[23]

In a new set of experiments, SDS-SUGE of protein mixtures was performed using a supramolecular hydrogel prepared from **1** (2.0 wt%), agarose (2.0 wt%), and TGS buffer (Tris: 25 mM;

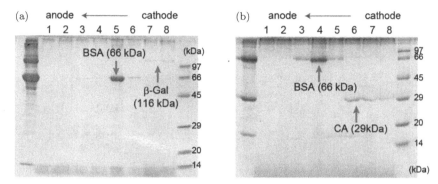

Fig. 3.5. SDS-PAGE analyses of SDS-SUGE separation using TGS buffer (25 mM Tris, 192 mM glycine, 1.0 mM SDS) of a) β-Gal (116 kDa) and BSA (66 kDa); b) BSA (66 kDa) and CA (29 kDa).

glycine: 192 mM; SDS: 1.0 mM) (Fig. 3.5). A mixture of β-Gal and BSA was separated by SDS-SUGE in the TGS buffer (25 mM Tris, 192 mM glycine, 1.0 mM SDS) at 100 V for 150 min (Fig. 3.5(a)). BSA (66 kDa), which is smaller than β-Gal, was detected in lanes 5 and 6, and the strongest band was found in lane 5. Larger β-Gal (116 kDa) was detected in lanes 7 and 8. SDS-SUGE of a mixture of BSA and OVA was performed at 100 V for 180 min. Larger BSA (66 kDa) was detected in lanes 4 and 5, with the strongest band found in lane 5. Smaller OVA (45 kDa) was detected in lanes 6 to 8, with the strongest band found in lane 7. OVA remained closer to the cathode than BSA. This result is considerably different from that obtained with SDS-SUGE using TGS buffer with 3.5 mM of SDS. Next, a mixture of BSA and CA was separated by SDS-SUGE at 100 V for 180 min (Fig. 3.5(b)). Larger BSA (66 kDa) was detected in lanes 3 to 5, with the strongest band found in lane 4, and smaller CA (29 kDa) was detected in lanes 6 to 8, with the strongest band found in lane 6. With this particular mixture, the smaller protein (CA) remained closer to the cathode than the larger protein (BSA). A mixture of OVA and CA also showed a similar separation, and the larger OVA was found closer to the anodic region than smaller CA. The threshold of separation using TGS buffer with 1.0 mM of SDS is therefore estimated to be about 66 kDa. Compared to SDS-SUGE using TGS buffer with 3.5 mM of SDS, the threshold for the different separation modes shifted towards a larger protein molecular weight.

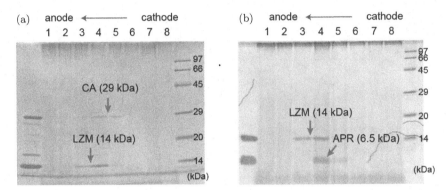

Fig. 3.6. SDS-PAGE analyses of SDS-SUGE separation using TGS buffer (25 mM Tris, 192 mM glycine, 7.0 mM SDS) of a) CA (29 kDa) and LZM (14 kDa); b) LZM (14 kDa) and APR (6.5 kDa).

A supramolecular hydrogel was prepared from tris-urea **1** (2.0 wt%), agarose (2.0 wt%), and TGS buffer (Tris: 25 mM; glycine: 192 mM; SDS: 7.0 mM), and was utilized for the SDS-SUGE of proteins (Fig. 3.6). A mixture of BSA and OVA was separated by SDS-SUGE in TGS buffer Tris: 25 mM; glycine: 192 mM; SDS: 7.0 mM) at 100 V for 180 min. Smaller OVA (45 kDa) showed a slightly larger mobility than larger BSA (66 kDa). The SDS-SUGE of a mixture of OVA and CA was performed at 100 V for 140 min. Larger OVA (45 kDa) was detected in lanes 4 to 7, and the strongest band was found in lane 5. Smaller CA (29 kDa) was detected in lanes 3 to 5, with the strongest band being found in lane 4. The SDS-SUGE of a mixture of CA and LZM was also performed at 100 V for 120 min (Fig. 3.6(a)). Larger CA (29 kDa) was detected in lanes 4 and 5, and smaller LZM (14.4 kDa) was detected in lanes 3 and 4. With these mixtures, the separation of proteins occurred in a similar way to that of traditional SDS-PAGE. When SDS-SUGE was run with a mixture of LZM and APR at 100 V for 120 min, larger LZM (14.4 kDa) was detected in lanes 3 and 4, while smaller APR (6.5 kDa) was detected in lanes 4 and 5 (Fig. 3.6(b)). Larger LZM demonstrated a larger mobility than smaller APR. When the TGS buffer is used with 7.0 mM of SDS, the threshold for different separation modes decreases to about 14 kDa. SDS-SUGE experiments using supramolecular hydrogel prepared from tris-urea **1** (2.0 wt%), agarose (2.0 wt%), and TGS buffer (25 mM Tris, 192 mM glycine, 6.0 mM SDS) display similar separation mechanisms to those of SDS-SUGE experiments with TGS buffer in 7.0 mM of SDS.

The threshold for the different separation mechanisms shifts with the concentration of SDS. A separation mechanism similar to that of gel filtration is dominant at low concentrations of SDS, and the threshold moves towards the large molecular weight range. In contrast, a separation mechanism analogous to that of SDS-PAGE is dominant at high concentrations of SDS, and the threshold appears in the small molecular weight range. The bundling ratio of **1** depends on the concentration of SDS; thinner fibers are observed on increasing the concentration of SDS. The size of the pores, which retain the proteins against the electric current, results from self-assembled aggregates of **1**. The pore size is smaller at higher concentrations of SDS than at lower concentrations of SDS.

3.3. Supramolecular Gel Electrophoresis (SUGE) for Native Proteins

3.3.1. *Development of a low molecular weight hydrogelator*[24]

Efficient recovery of electrophoresed proteins is an important task in electrophoresis of native proteins. However, strong interaction between the polyacrylamide gel and native proteins remains a hindrance in native-PAGE, a widely used technique. Efficient recovery of electrophoresed proteins was achieved using the above-mentioned SDS-SUGE method. This result encouraged us to develop a SUGE technique for native proteins as an alternative technique of native-PAGE.

Unfortunately, amphiphilic tris-urea **1** could not form supramolecular hydrogel with Tris-glycine (TG) buffer (Tris: 25 mM; Glycine: 192 mM), which is a typical buffer used for native-PAGE. Accordingly, we first had to develop a novel low molecular weight hydrogelator for TG buffer. More hydrophilic structures than **1** seemed to be suitable for a gelator of TG buffer, and amphiphilic tris-urea **2** was designed as a low molecular weight hydrogelator for TG buffer. Hydrophobic moieties of **1** were reduced in **2**. Synthesis of **2** was

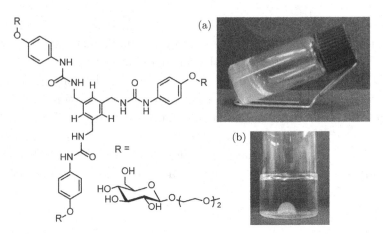

Fig. 3.7. Chemical structure of amphiphilic tris-urea **2** and photographs of a) hydrogel of **2** (0.25 wt%); b) hydrogel of **2** (1.0 wt%) in water.

achieved by a multistep reaction started from trimesic acid and α-D-glucose. Supramolecular hydrogels of **2** were formed in not only TG buffer but also various aqueous solutions (Fig. 3.7).[24]

A mixture of **2** (0.25 wt% = 1.9 mM) and H_2O afforded a semi-transparent supramolecular hydrogel after thermal dissolution of the mixture (Fig. 3.7(a)). An opaque hydrogel was obtained from a higher concentration of **2** (1.0 wt%) in H_2O. Once formed, the hydrogels were stable for at least a year at ambient temperature, regardless of the concentration of **1**. Moreover, a lump of the hydrogel of **2** (1.0 wt%) kept its shape for a year, even in water, without dispersion or swelling (Fig. 3.7b). These hydrogels exhibited thermoresponsive behavior, and reversible gel−sol phase transitions resulted from heating and cooling mixtures. Furthermore, these hydrogels exhibited thixotropy. A low-viscosity liquid was obtained after shaking the hydrogel of **2** using a vortex vibrator for 5 sec. The hydrogel was recovered from the mechanoinduced liquid state within 10 min. The mechanoresponsive gel−sol phase transition was repeatable for more than 30 times.

The amphiphilic tris-urea **2** formed hydrogels with physiological saline, simulated body fluid (SBF), and seawater (from Suruga Bay). Generally, supramolecular gels sensitively responded to anions, and showed gel−sol phase transition.[25] Physiological saline contains about 80 equiv of NaCl for 0.25 wt% of **2**. Even saturated solutions of NaF (0.84 M) and NaCl (5.35 M) are readily gelled by 0.25 wt% of **2**. The tolerable concentration of NaBr was estimated to be 3.8 M (2000 equiv for **2**) to form a gel by 0.25 wt% of **2**; however, saturated solutions of NaBr (7.12 M) were gelled by 1.0 wt% of **2**. The gelation of **2** was slightly inhibited by NaI. A partial gel or opaque sol was obtained from a mixture of **2** (0.25 wt%) and 0.76 M (400 equiv for **2**) or 1.33 M (700 equiv for **2**) aqueous solutions of NaI, respectively. The same tendency was observed in gelation experiments of **2** in the presence of ammonium salts.

The amphiphilic tris-urea **2** formed hydrogels with a wide range of pH solutions. A mixture of **2** (1.0 wt%) and 0.01 M HCl (pH = 2.0) gave an opaque hydrogel. Even 8 M HCl formed a hydrogel in the presence of 4.0 wt% of **2**. Although, hydrolysis of acetal moieties of **2** proceeded during hydrogelation with acidic solutions, resulting supramolecular hydrogels were stable. A hydrogel was formed in the solution of 0.01 M NaOH (pH = 12.0) by using 1.0 wt% of **2**. Even the much stronger alkaline 7 M KOH solution was gelled by 2.0 wt% of **2**.

The amphiphilic tris-urea **2** formed supramolecular hydrogels with various types of buffers. It is an important characteristic for biological application. For example, glycine–IICl buffer (pII = 3.0), phosphate–NaOH buffer (pH = 6.9), HEPES [4-(2-hydroxyethyl)-1-piperazineethanesulfonic acid] buffer (pH = 7.0), tris–HCl buffer (pH = 7.4), borate–NaOH buffer (pH = 9.1) formed hydrogels in the presence of 0.25 wt% of **2**. A hydrogel was also obtained from a mixture of **2** (0.25 wt%) and DMEM (Dulbecco's modified Eagle medium), a typical medium for cell culture.

3.3.2. *Development of Native-SUGE*[26]

Electrophoresis of native proteins using supramolecular hydrogel of **2** was performed using the following procedure, which is similar to SDS-SUGE (Fig. 3.2). 1) A supramolecular hydrogel was prepared from amphiphilic tris-urea **2** (2.0 wt%) and a TG buffer (Tris: 25 mM; glycine: 192 mM; pH: 8.3). 2) A glass capillary (with ϕ 2 mm, length 120 mm) was filled to 80 mm with the supramolecular hydrogel. 3) Sample proteins solution was adsorbed onto one end of the hydrogel, and both ends of the capillary were filled with agarose gel (2.0 wt%). 4) The glass capillary was immersed in TG buffer (Tris: 25 mM; glycine: 192 mM; pH: 8.3) in a submarine electrophoresis system and electrophoresed at 100 V for 100 min. 5) The electrophoresed gel was then removed from the glass capillary and divided into eight equal parts (numbered 1–8 from the anode), and proteins were isolated from the supramolecular hydrogel by freezing and centrifugation. The supernatant derived from this procedure contained proteins in their native states. 6) The separation pattern was performed by analyzing the supernatants using a standard SDS-PAGE system followed by CBB staining.

Three acidic native proteins, D-lactate dehydrogenase (LDH, tetramer = 146 kDa, pI = 4.0), β-galactosidase (β-Gal, tetramer = 540 kDa, pI = 4.6), and ovalbumin (OVA, monomer = 45 kDa, pI = 4.7), were used in the initial experiments of native-SUGE using the supramolecular hydrogel of **2**. Two of the three proteins were applied to the native-SUGE system and their separation patterns were analyzed by SDS-PAGE (Fig. 3.8).

First, a mixture of LDH and β-Gal was employed for native-SUGE (Fig. 3.8(a)). LDH was found in lanes 4 and 5, with lane 4 containing the strongest band. β-Gal was found in lanes 5 and 6, with lane 5 containing the strongest band. The smaller (146 kDa) and more acidic (pI = 4.0) LDH was therefore more mobile than the larger (540 kDa) and less acidic (pI = 4.6) β-Gal in this system. LDH and β-Gal retained their enzymatic activities after electrophoresis in the supramolecular hydrogel of **2**, indicating that these proteins remained in their native tetrameric forms. LDH and β-Gal were denatured in the process of performing SDS-PAGE after electrophoresis in the supramolecular hydrogel of **2** had been performed; thus, LDH and β-Gal were found to have molecular weights of 36.5 kDa and 135 kDa, respectively, by SDS-PAGE analysis. Next, we used a mixture of LDH and OVA to perform native-SUGE (Fig. 3.8(b)). In this experiment, LDH was found in lanes 3 and 4, with lane 4 containing the strongest band, and OVA was found in lanes 5 and 6, with lane 5 containing the strongest band. In this experiment, the larger and more acidic LDH was more mobile than the smaller and less

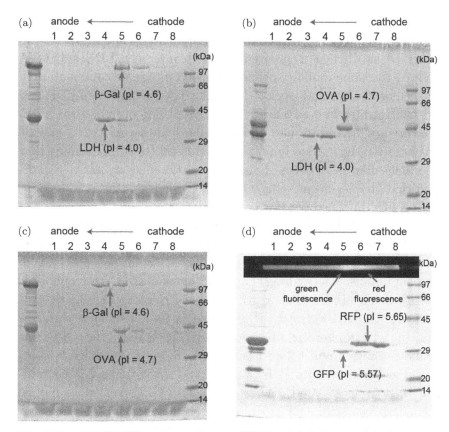

Fig. 3.8. SDS-PAGE analyses of native-SUGE of a) D-lactate dehydrogenase (LDH, 146 kDa, pI = 4.0) and β-galactosidase (β-Gal, 540 kDa, pI = 4.6); b) LDH and ovalbumin (OVA, 45 kDa, pI = 4.7); c) β-Gal and OVA; d) green fluorescent protein (GFP, 27 kDa, pI = 5.57) and red fluorescent protein (RFP, 27 kDa, pI = 5.65) and photographs of the UV irradiated glass capillary after electrophoresis (insert).

acidic OVA. A mixture of β-Gal and OVA was next used to perform native-SUGE (Fig. 3.8(c)). SDS-PAGE analysis showed β-Gal bands of similar densities in lanes 4 and 5. OVA bands were found in lanes 5 and 6, with lane 5 containing the strongest band. In this experiment, the larger and more acidic β-Gal was more mobile than the smaller and less acidic OVA.

These results indicate that native-SUGE using the supramolecular hydrogel of **2** caused acidic native proteins to be separated depending on their pI values rather than their molecular weights.

The typical native-PAGE experiments of these proteins showed considerably different results. LDH and β-Gal, and β-Gal and OVA were observed as distinct, separated bands. Conversely, LDH and OVA showed the same mobility owing to their comparable combination of molecular weights and isoelectric points. The separation of LDH and OVA illustrates one advantage of native-SUGE over native-PAGE. The meshes present in the supramolecular hydrogel of **2** (2%) may be coarser than the meshes in polyacrylamide gel (>10%), meaning that the molecular sieve effect should have little effect in native-SUGE, and that the net negative charge of the proteins is the predominant factor affecting their separation.

Native-SUGE was performed using green fluorescent protein (GFP, 27 kDa, pI = 5.57) and red fluorescent protein (RFP, 27 kDa, pI = 5.65). The green and red fluorescent band was observed in the supramolecular hydrogel of **2** during and after electrophoresis via ultraviolet (UV, 365 nm) irradiation of the gel (Fig. 3.8d). This indicates that GFP and RFP remained in its native form throughout the electrophoretic process. After SDS-PAGE, the GFP was found in lanes 5 and 6 (with lane 5 containing the strongest band) and the RFP was found in lanes 6 and 7 (with lane 7 containing the strongest band). The more acidic GFP was more mobile than the less acidic RFP, despite their having similar isoelectric points.

It is preferable for proteins to retain their native 3D structures and activities during electrophoresis. The activity of the LDH after the native-SUGE was measured to confirm that it had remained in its native form. LDH catalyzes the oxidation of lactate to pyruvate in the presence of nicotinamide adenine dinucleotide (NAD^+). The enzymatic activity of LDH can be determined by the absorption of UV light at 340 nm by the reduced form, NADH. The LDH activity after electrophoresis through the supramolecular hydrogel of **2** was determined using D-/L-lactic acid assay kit (Biocon Japan Ltd.), according to the manufacturer's instructions. LDH (6.0 μg) was electrophoresed through the supramolecular hydrogel of **2** (2.0 wt%, ϕ = 2 mm, length 20 mm) at 100 V for 30 min, and the whole gel was used for the assay. The LDH activity after electrophoresis was around 97% of the original LDH activity.

3.3.3. *Affinity electrophoresis of lectin*[26]

Carbohydrate-binding proteins (lectins) have many important roles in living organisms, including information transmission.[27] Affinity electrophoretic systems for lectins have been developed and are effective for identifying protein functions.[6] The surfaces of the nanofibers formed through the self-assembly of the amphiphilic tris-urea **2** were densely coated with glucosides that were introduced as hydrophilic groups. The glucoside-coated nanofibers can interact with appropriate lectins during electrophoresis, and the lectins involved in these interactions do not migrate as far as non-carbohydrate-binding proteins in the gel. A well-characterized lectin, concanavalin A (ConA, tetramer = 112 kDa, pI = 4.4–5.5), was subjected to affinity electrophoresis using the supramolecular hydrogel of **2**. A moderate association between **2** and ConA was expected, from the known affinity between α-methyl-D-glucopyranoside and ConA (K_a = 1.96 $\times$ 10^3 M^{-1}).[28]

The electrophoresis of ConA was performed under typical native-SUGE conditions (2.0 wt% of **2**, 100 V, 100 min) (Fig. 3.9). SDS-PAGE showed that most of the ConA remained at the cathode at the starting point (lane 8), and the ConA remained in lane 8 even when the electrophoretic time was increased to 300 min (Fig. 3.9a). In contrast, the electrophoresis of denatured-ConA under the same conditions showed considerably different results, with the denatured-ConA migrating towards the anode during electrophoresis using the supramolecular hydrogel of **2**, and was found in lanes 4 and 5 after SDS-PAGE (Fig. 3.9b). These results indicate that native ConA interacted with the glucosides on the nanofiber surfaces, which inhibited its electrophoretic mobility. The addition of a saccharide with a strong affinity for ConA to the electrophoresis buffer may improve the electrophoretic migration of native ConA in the native-SUGE. The ConA and saccharide complex in the buffer prevents ConA interacting with the glucosides on the nanofiber surfaces and allows ConA to migrate towards the anode during electrophoresis. The saccharide α-methyl-D-mannopyranoside (MeαMan) was selected for this task because it has strong affinity for ConA (K_a = 0.82 $\times$ 10^4 M^{-1}). A TG–MeαMan buffer (25 mM Tris, 192 mM glycine, 51 mM MeαMan,

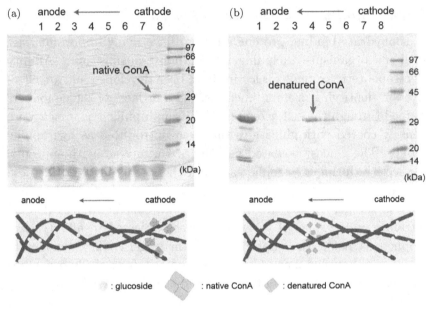

Fig. 3.9. SDS-PAGE analyses of native-SUGE of a) concanavalin A (ConA, 112 kDa (as tetramer), pI = 4.4–5.5); b) denatured ConA.

pH = 8.3) was used in native-SUGE for native ConA. After SDS-PAGE analysis, ConA was found in lanes 5 and 6. The native ConA migrated much further toward the anode in the presence of MeαMan than it did under typical native-SUGE conditions.

3.4. Conclusion

Protein electrophoresis is an indispensable experimental technique for research in the life sciences. There are many existing methods for protein electrophoresis; however, electrophoresis using polyacrylamide gel has been an established technique for several decades. The creation of a novel gel carrier is key to continuing the development of protein electrophoresis. We have focused on supramolecular hydrogel, with highly flexible characteristics, as a novel carrier, and attained electrophoreses for denatured and native proteins (SDS-SUGE and native-SUGE). In SDS-SUGE, a unique separation manner appeared on account of two different molecular sieve effects. In native-SUGE, proteins were separated on the basis of isoelectric points of proteins rather than their molecular weights. Protein samples were recovered efficiently by extremely simple procedures in either types of SUGE. Further investigations may lead to SUGE becoming an innovative experimental technique in life sciences research.

References

1. Kurien, B. T. and Scofield, R. H. (2012). *Protein Electrophoresis Methods and Protocols* (Springer Science).
2. Tiselius, A. (1937). Electrophoresis of serum globulin. I, *Biochem. J.* **31**, 313–317.
3. Smithies, O. (1955). Grouped variations in the occurrence of new protein components in normal human serum, *Nature* **175**, 307–308.
4. Raymond, S. and Weintraub, L. (1959). Acrylamide gel as a supporting medium for zone electrophoresis, *Science* **130**, 711.
5. Laemmli, U. K. (1970). Cleavage of structural proteins during the assembly of the head of bacteriophage T4, *Nature* **227**, 680–685.
6. Takeo, K. (1995). Advances in affinity electrophoresis, *J. Chromatogr. A* **698**, 89–105.
7. Görg, A., Weiss, W. and Dunn, M. J. (2004). Current two-dimensional electrophoresis technology for proteomics, *Proteomics* **4**, 3665–3685.
8. Kinoshita, E., Kinoshita-Kikuta, E. and Koike, T. (2009). Separation and detection of large phosphoproteins using Phos-tag SDS-PAGE, *Nature Protocol* **4**, 1513–1521.
9. Hiratsuka, A. *et al.* (2007). Fully automated two-dimensional electrophoresis system for high-throughput protein analysis, *Anal. Chem.* **79**, 5730–5739.
10. Terech, P. and Weiss, R. G. (1997). Low molecular mass gelators of organic liquids and the properties of their gels, *Chem. Rev.* **97**, 3133–3159.
11. Estroff, L. A. and Hamilton, A. D. (2004). Water gelation by small organic molecules, *Chem. Rev.* **104**, 1201–1218.
12. de Loos, M., Feringa, B. L. and van Esch, J. H. (2005). Design and application of self-assembled low molecular weight hydrogels, *Eur. J. Org. Chem.* 3615–3631.
13. Buerkle, L. E. and Rowan, S. J. (2012). Supramolecular gels formed from multi-component low molecular weight species, *Chem. Soc. Rev.* **41**, 6089–6102.
14. Babu, S. S., Praveen, V. K. and Ajayaghosh, A. (2014). Functional π-gelators and their applications, *Chem. Rev.* **114**, 1973–2129.
15. Hirst, A. R., Escuder, B., Miravet, J. F. and Smith, D. K. (2008). High-tech applications of self-assembling supramolecular nanostructured gel-phase materials: From regenerative medicine to electronic devices, *Angew. Chem. Int. Ed.* **47**, 8002–8018.

16. Yang, Z., Liang, G. and Xu, B. (2008). Enzymatic hydrogelation of small molecules, *Acc. Chem. Res.* **41**, 315–326.
17. Cui, H., Webber, M. J. and Stupp, S. I. (2010). Self-assembly of peptide amphiphiles: From molecules to nanostructures to biomaterials, *Biopolymers* **94**, 1–18.
18. Ikeda, M., Ochi, R. and Hamachi, I. (2010). Supramolecular hydrogel-based protein and chemosensor array, *Lab Chip* **10**, 3325–3334.
19. Yamamichi, S., Jinno, Y., Haraya, N., Oyoshi, T., Tomitori, H., Kashiwagi, K. and Yamanaka, M. (2011). Separation of proteins using supramolecular gel electrophoresis, *Chem. Commun.* **47**, 10344–10346.
20. Yamanaka, M., Nakamura, T., Nakagawa, T. and Itagaki, H. (2007). Reversible sol-gel transition of a tris-urea gelator that responds to chemical stimuli, *Tetrahedron Lett.* **48**, 8990–8993.
21. Yamanaka, M. (2016). Development of C_3-symmetric tris-urea low-molecular-weight gelators, *Chem. Rec.* **16**, 768–782.
22. Jinno Y. and Yamanaka, M. (2012). Ionic surfactants induce amphiphilic tris-urea hydrogel formation, *Chem. Asian. J.* **7**, 1768–1771.
23. Tazawa, S., Kobayashi, K. and Yamanaka, M. (2016). Effect of sodium dodecyl sulfate concentration on supramolecular gel electrophoresis, *Chem. Nano. Mat.* **2**, 423–425.
24. Higashi, D., Yoshida, M. and Yamanaka, M. (2013). Thixotropic hydrogel formation in various aqueous solutions through self-assembly of an amphiphilic tris-urea, *Chem. Asian. J.* **8**, 2584–2587.
25. Haraya, N., Yamamichi S. and Yamanaka, M. (2011). Chemical stimuli-responsive supramolecular hydrogel from amphiphilic tris-urea, *Chem. Asian J.* **6**, 1022–1025.
26. Munenobu, K., Hase, T., Oyoshi, T. and Yamanaka, M. (2014). Supramolecular gel electrophoresis of acidic native proteins, *Anal. Chem.* **86**, 9924–9929.
27. Lis, H. and Sharon, N. (1998). Lectins: Carbohydrate-specific proteins that mediate cellular recognition, *Chem. Rev.* **98**, 637–674.
28. Dam, T. K. and Brewer, C. F. (2002). Thermodynamic studies of lectin–carbohydrate interactions by isothermal titration calorimetry, *Chem. Rev.* **102**, 387–429.

CHAPTER 4

Self-assembly of Polymer-grafted Inorganic Nanoparticles into Functional Hybrid Materials

CHENGLIN YI and ZHIHONG NIE*

Department of Chemistry and Biochemistry, University of Maryland
College Park, MD 20742, USA

4.1. Introduction

Current interest in functional assemblies of inorganic nanoparticles (NPs) is stemmed from their collective properties that may vastly differ from discrete NPs and corresponding bulk materials. These promising properties mainly arise from coupling interactions such as, plasmon-plasmon, plasmon-exciton, and magnetic-magnetic interactions between the NP building blocks.[1] Although breakthroughs in the synthesis of a wide spectrum of inorganic NPs with different sizes, shapes, compositions and morphologies have been achieved over the past few decades, controlled assembly of inorganic NPs into architectures with desired complexity and functionality remains

---
*Corresponding author: znie@umd.edu

a grant challenge. Generally, this challenge can be met by modifying the surface of inorganic NPs with organic coatings to tailor the interactions between NPs and hence their organization in space.[2] To date, a spectacular variety of small molecular, biomolecular, and polymeric coatings have been employed for this purpose.[2-5] Among others, biomolecular coatings (e.g., DNA) are distinguished by their highly biospecific interactions. In contrast, polymer coatings feature fascinating diversity in the molecular structures, architecture and functionalities that polymers can offer. This has been fueled by recent advances in polymer chemistry along with long-standing research interest in block copolymer (BCP) self-assembly.[6] The anchoring of polymers on inorganic NP surfaces generates a new class of hybrid building blocks, namely "hairy" inorganic nanoparticles (HINPs). HINPs intrinsically combine the unique features of both inorganic NP cores and tethered polymers.[7,8]

Organization of HINPs into complex structures can often draw inspirations from principles underlying atomic packing and molecular self-assembly.[9-12] Recent advances in this frontier have brought recognition that anisotropic interactions through chemical 'patches' is crucial to the self-assembly of HINPs into desired assembly structures.[9,10] Regiospecific surface modifications are usually used to decorate inorganic NPs with surface patches of polymers or other molecules. Common strategies can be generally summarized into two categories: (1) Selective attachment of polymers to certain crystal facets of NPs due to differential binding strength of functional groups to crystal facets.[13-15] For example, thiol-terminated polystyrene (PS) can preferentially attach to both tips of gold nanorods (AuNRs), leading to the formation of two reactive polymeric patches at both ends of AuNRs.[13,16] When they are exposed to selective solvents with respect to PS, the association between the hairy polymers on AuNRs can drive the assembly of AuNRs into a variety of assemblies, such as chains, bundles, rings, and vesicles. Another version of this strategy is the attachment of different polymers onto different domains of multicomponent inorganic NPs. As an example, polyethylene glycol (PEO) and PS can selectively bind to different components of Janus Au-Fe_3O_4 NPs to produce Janus HINPs with hydrophobic and hydrophilic halves.[15] The resulting building blocks resemble molecular amphiphiles to assemble into double layer plasmonic-magnetic vesicles. (2) Localized grafting of polymers by partially masking or protecting selective surface regions of NPs.[17] Generally, solid-liquid or liquid-liquid interfaces are used to selectively mask part of the surface of NPs, while the rest of unprotected surface is prone to post-modification to generate Janus-like or more complex HINPs.[17] The 'masking strategy' has been widely used for functionalizing microsized colloidal particles, for example, to construct molecular-like supracolloids.[18] This method does not offer sufficient resolution for selective surface functionalization of nanosized particles (particularly for NPs with diameter below ~100 nm) due to the relatively low resolution of the masking edges. Nevertheless, despite growing interest in the self-assembly of HINPs decorated with multi-compartment brushes, anisotropic functionalization and surface characterization of resulting NPs remain a grant challenge.

Interestingly, recent compelling evidence suggests that anisotropic functionalization may not be necessary for triggering the precise or hierarchical assembly of HINPs and highly directional interactions may arise between NPs isotropically decorated with polymer.[19,20] The relative simplicity of isotropic functionalization makes this class of HINPs particularly intriguing from the perspectives of cost, scaling-up, and processing. To date, there exist few reviews summarizing the development of HINPs from different perspectives.[4,7,21-25] This chapter provides a review on the recent progress in the self-assembly of HINPs carrying isotropically functionalized polymers, with emphasis on the structure–property relationships of HINPs.

4.2. Structural Parameters of HINPs

The surface functionality and interparticle interactions of HINPs are largely dependent on the conformation of polymer brushes on NP surfaces, which is effected by grafting density (σ), polymer length and composition, and particle size and shape. Take spherical HINPs dispersed in good solvents as an example, the brush conformation could be categorized into three different regimes (Fig. 4.1).[7,26] At low $\sigma (\sigma < R_g^{-2})$, where R_g is the radius of gyration of the chains),

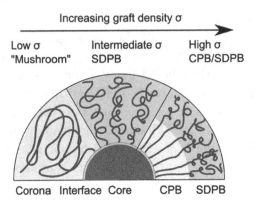

Increasing graft density σ

Low σ	Intermediate σ	High σ
"Mushroom"	SDPB	CPB/SDPB

Corona Interface Core   CPB   SDPB

Fig. 4.1. Schematic illustration of the three reigmes of chain confirmation for polymer brushes grafted on the surface of NPs. Reprinted with permission from Ref. 7. Copyright 2013 Material Research Society.

polymer chains adopt an approximately random coil conformation; that is the *so-called* 'mushroom' regime. As the σ exceeds the chain overlap threshold, polymers are stretched from the anchor points towards solvent media, resulting in a transition to the semi-dilute polymer brush (SDPB) regime. At an even higher σ, a concentrated polymer brush (CPB) regime is accessed, which is featured by non-Gaussian chain characteristics and more extended chain conformations. For spherical surfaces symmetrically grafted with polymers, the associated scaling laws for CPB regime and SDPB regime are $h \propto N^{0.8}$ and $h \propto N^{0.6}$, where h is the brush height and N is the degree of polymerization for the grafted polymers.[26] It is worth noting that the tip of polymers grafted on spherical HINPs preferentially adopts a SDPB conformation due to the increase in the interchain spacing with increasing distance away from the surface, even when the root segments are in CPB regime.

4.3. Tailored Synthesis of HINPs

(1) 'One-pot' synthesis

Current strategies for the preparation of HINPs generally fall into three categories, namely one pot synthesis, and "grating from" and "grafting to" methods. In the former method, the NP synthesis and surface modification occur simultaneously in one reaction system in which polymers such as polyvinylpyrrolidone (PVP) are used as both stabilizer and shape directing agent for the growth of inorganic NPs (e.g., Au or Ag NPs). The stability of resulting HINPs relies on the strong affinity between functional groups on polymers and NP surfaces. Organic solvents (e.g. DMF, THF, or toluene) are often used for the synthesis of inorganic NPs in the presence of hydrophobic polymer ligands.[27] Moreover, BCP micelles[28] or unimolecular micelles[29,30] can serve as nanoreactors for the synthesis of monodisperse HINPs with tunable size by enriching functional groups with strong affinity to metal precursors in the confined space. When conventional BCP micelles are used, the synthesis often produces HINPs with many domains of inorganic NPs stabilized by the corona of polymer micelles. In contrast, unimolecular micelle-templated synthesis offers much better control over the uniformity, crystallinity, size, shape and composition of resulting inorganic NPs. This strategy requires high-precision control over structural parameters of polymer architectures. For example, Zhiqun *et al.* recently developed several types of bottlebrush BCPs as nanoreactors for the growth of hairy NPs with controlled size, shape (e.g., nanospheres and nanorods), and compositions.[29,30] Figure 4.2 shows the synthesis of 1D nanocrystals with plain, core-shell and hollow structures by using precisely designed nanoreactors of warm-like micelles. The rational design on the structural parameters (e.g., the composition and length of each polymer segments, as well as the σ of brushes) of bottlebrush polymers is critical to the formation of inorganic NPs with high crystallinity. Take core-shell NPs as an example, the sequential deposition of different inorganic materials requires that the middle block of poly(*tert*-butyl acrylate) (P*t*BA) brushes can not only selectively protect the growth of Au corebut also act as nano-reactor for the secondary deposition of metal oxides, after the hydrolysis of *tert*-butyl acrylate in P*t*BA into acrylic acid (Figs. 4.2(b), (e)).

(2) 'Grafting from'

Both "grafting from" and "grafting to" methods use post surface modification of as-synthesized NPs to produce HINPs. Recent advances in the synthesis of inorganic NPs with different sizes, shapes and compositions significantly enlarge the library of HINPs that can

(a)

(b)

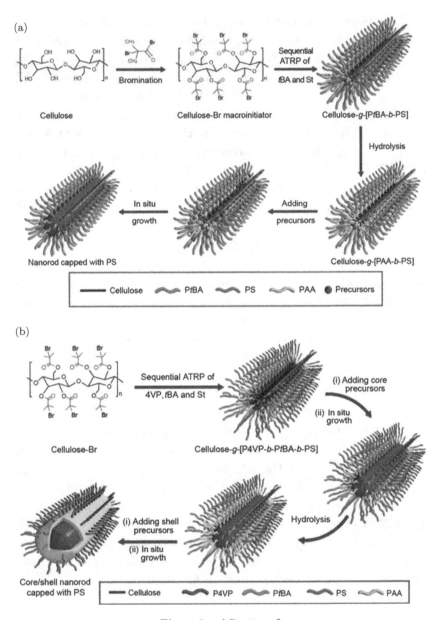

Fig. 4.2. (*Continued*)

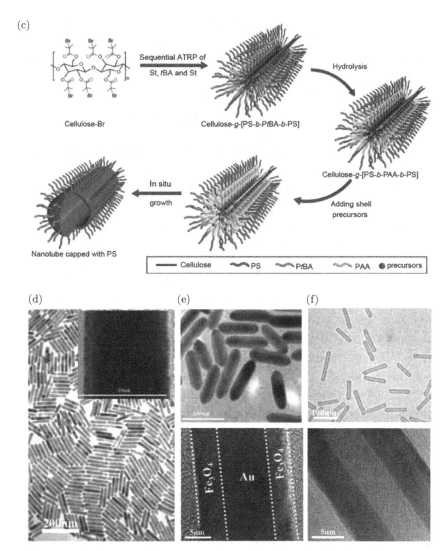

Fig. 4.2. (a–c) Schematics illustrating the synthesis of hairy nanorods by using cylindrical bottlebrushes polymers as nanoreactors: (a) plain nanorods templated by cellulose-*g*-[poly(acrylic acid)-*b*- poly(styrene)] (PAA-*b*-PS). St, styrene; *t*BA, *tert*-butyl acrylate; (b) core-shell nanorods templated by cellulose-*g*-[poly(4-vinylpyridine)-*b*- poly*(tert*-butyl acrylate)-*b*-poly(styrene)] (P4VP-*b*-P*t*BA-*b*-PS); (c) nanotubes templated by cellulose-*g*-(PS-*b*-PAA-*b*-PS). (d–f) Representative TEM images of three typical hairy NRs: (d) core Au NRs synthesized by using nanoreactor cellulose-*g*-(P4VP-*b*-P*t*BA-*b*-PS), (e) Au-Fe$_3$O$_4$ nanorods (Fe$_3$O$_4$ shell thickness $t = 4.6 \pm 0.4$ nm), and (f) Au nanotubes ($L = 103 \pm 12$ nm, $t = 5.1 \pm 0.5$ nm, hollow interior $D = 5.3 \pm 0.4$ nm). Reprinted with permission from Ref. 29.

be made for use as building blocks in assembly. "Grafting from" method takes advantage of surface-initiated controlled/"living" polymerization, such as Atom Transfer Radical Polymerization (SI-ATRP) to grow polymers from NP surfaces.[31] The resulting HINPs feature high σ (up to $4 \sim 5$ chains/nm^2), due to the high diffusivity of initiators/monomers compared to polymers. To obtain high quality HINPs, several potential problems that need to be solved include such as, NP stability during polymerization and cross-linking of growing polymers caused by locally concentrated free radicals. For example, immobilization of ATRP initiators and polymers (e.g., PEO) with appropriate ratio on NP surfaces can improve the stability of NPs and prevent the aggregation of NPs during polymerization.[32,33]

Moreover, by combining two or more polymerization methods, 'grafting from' strategy can be used produce various HINPs with binary brushes with tailorable structural parameters. For example, Bin Zhao *et al.* established a general approach to fabricate mixed brushes grafted NPs by immobilizing 'Y' initiators on the surface of SiO$_2$ NPs and using sequential surface-initiated controlled/"living" polymerizations.[34–39] Their method features the highest grafting homogeneity. Almost equivalent amount of binary brushes can be grown from NP surface. Meanwhile, the relatively high σ also ensures the isotropic attachment of polymers on NP surface to obtain HINPs with controlled relatively size ratio of polymers to inorganic core (Fig. 4.3(a)). Figure 4.3(b) shows that microphase-separations between homogeneously mixed binary brushes could readily occur and form 'rippled' structures on the surface of inorganic core at a relative large size ratio of polymer brushes to the NP core.[36] The periodicity (D) of the 'rippled' structures was found to be logarithmically linearly increased with the increase in the molecular weight (MW) of polymer brushes (Fig. 4.3(c)).

(3) 'Grafting to'

In "grafting to" method, as-synthesized polymers are grafted onto the surface of NPs through chemical reaction or ligand exchange. Polymers used for this purpose must contain one or more functional groups that can be either attached to original capping ligands on NP

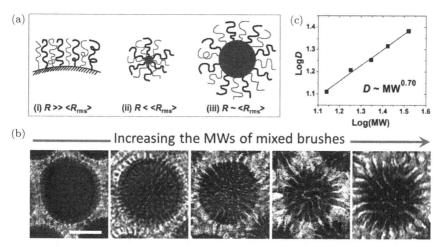

Fig. 4.3. (a) Mixed homopolymer brushes grafted on spherical particles with different relative sizes of radius R to root-mean-square end-to-end distances ($\langle R_{\mathrm{rms}} \rangle$) of polymers. Reprinted with permission from Ref. 34. (b) Effect of MW of mixed brushes on their lateral microphase separation on the surface of SiO_2 NPs. From left to right: the average MW of binary brushes increasing from 13.8 kDa to 33.1 kDa. Scale bar is 100 nm. The samples were stained with RuO_4 at room temperature for 20 min. (c) Plot of $\log D$ versus $\log(\text{MW})$, where D is the average ripple periodicity and MW is the average MW of PtBA and PS in the unit of kDa. The straight solid line is a linear fit with $R = 0.997$ and a slope of 0.70. Reprinted with permission from Ref. 36.

surface,[40] or directly anchored onto the surface of inorganic NP.[41–43] A recent review by Mattoussi *et al.*[44] offers an excellent summary of ligand exchange strategies and functional groups that can anchor on specific inorganic substrate. Despite its simplicity and mild experimental condition, this approach usually generates HINPs with relatively low grafting densities, due to the steric hindrance of polymers on NP surface. The already attached polymers create a steric barrier for free polymers to reach the binding cites on NP surfaces. This effect is dominant particularly for longer polymers.[45] Moreover, the presence of small ligands originated from as-synthesized NPs influences the attachment of polymers; on the other hand, it also provides a route to tune σ of polymers attached onto the NPs.[46,47]

With regard to σ and homogeneity, the two post-modification methods show their own pros and cons. In general, the 'grafting from' approach allows for achieving a high polymer σ on NPs. It, however, shows possible limitations, such as relatively complex synthetic procedures, crosslinking of polymers and even NPs, and difficulty in scale up. In contrast, the "grafting to" strategy leads to a more uniform polymer coverage on the surface of inorganic NPs at a given σ. As aforementioned, the excluded volume constraints of the previously occupied chains limit the polymer σ at the low-to-moderate regime.[48]

It is interesting that relatively low σ of polymers on HINPs can be beneficial to the assembly of HINPs and the lateral phase-separation of the polymer brushes on the surface of NPs.[20,46] As an example, Kumacheva *et al.* demonstrated that the surface patterning of polymer brushes on NPs at a relatively low σ of polymers on Au NPs with different shapes.[20] The process led to the generation of NPs with two or more polymer patches, as well as NPs with surface ripples[36] or 'raspberry' morphology. The polymer patterning on the NP surfaces relies on the thermodynamically driven segregation of polymer ligands from a uniform polymer brush into surface-pinned micelles upon the change in solvent quality. Figure 4.4 shows that the number of pinned micelles formed on NPs is dependent on the size and shape of NP core, and the length and σ of polymer brushes. It is worth noting that pinned micelles preferentially locate at the tips and edges of shaped NPs, due to the high surface energy of those regions. A phase-diagram further emphasizes on the effect of σ and relative size of NPs on the phase-segregation and pattern formation of polymer brushes on the surface of NPs (Fig. 4.4(d)). This work provides us a clear vision on the adaptive change of polymer confirmation on highly curve, nanoscale surface in response to selective solvents, as well as the potential of hairy NPs as novel building blocks in contrasting functional materials.

4.4. Anisotropic Self-assembly of HINPs

The anisotropic assembly of HINPs could be assisted by external fields, such as, magnetic field and *d.c.* or *a.c.* electric field.[49] During this process, polymer brushes serve as steric stabilizers, while the assembly is triggered by the permanent or induced dipole interactions of inorganic NP cores under field. Our discussion here is mainly

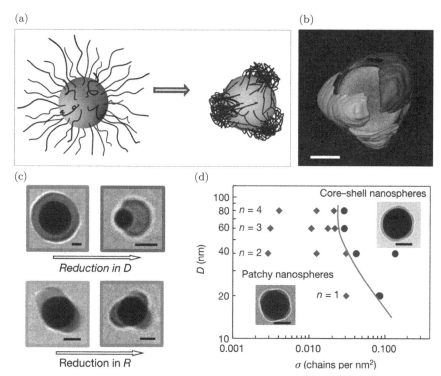

Fig. 4.4. (a) Schematics of solvent-mediated formation of pinned polymer micelles (surface patches) on the nanoparticle surface. (b) Electron tomography reconstruction image of Au nanosphere with three PS patches. PS brushes have a MW of 50 kDa and σ of 0.02 chains/nm². Scale bar is 20 nm. (c) Effect of Au NP size (top) and polymer dimensions (bottom) on patch formation. The nanospheres are functionalized with polystyrene-50K (top row and bottom left) and polystyrene-30K (bottom right) at $\sigma = 0.03$ chains/nm². Scale bars are 25 nm. (d) Phase diagram for the patterning of polymer brushes on AuNPs. The blue line indicates the boundary between regions forming core–shell and patchy nanospheres of different average patch number n. Scale bars are 50 nm. Reprinted with permission from Ref. 20.

focused on the role of grafted polymer brushes in engineering the interparticle interactions and hence their organization into structured materials. Specifically, we will summarize recent efforts on designing the chemical and/or shape anisotropy of HINPs for assembling polymer nanocomposites in both the condensed state and in aqueous solution.

4.4.1. *Self-assembly of HINPs in condensed state*

Although HINP self-assembly resembles that of organic molecules, the mobility and relaxation of HINPs is significantly slower than that of organic molecules, due to their large dimensions.[50] This makes the assembly kinetics and thermodynamics of HINPs vastly distinct from molecular systems.

4.4.1.1. *Assembly at the liquid/Air interface*

The liquid/air interfaces can be served as template to direct the self-assembly of HINPs into various 2D and 3D lattice structures. The lateral capillary forces and the reduction of interfacial energy is the dominant driving forces for the interfacial assembly of HINPs. For example, Chen *et al.* recently reported the evaporation-mediated assembly of PS-modified silver nanocubes (AgNCs) into free-standing superlattice monolayers at the air-water interface (Figs. 4.5(a, b)).[51] The evaporation of a dispersion of such HINPs in chloroform on the surface of water produced 2D hexagonal NP arrays with long-range order and good stability. The AgNC sheets can be manufactured into nanoribbons or origami using focused ion beams (Fig. 4.5(c)). At high concentrations of HINPs with polymer conformation in the CPB/SDPB regime, their self-assembly into NP superlattices is usually governed by excluded volume effect associated with the interpenetration of polymer brushes on neighboring NP surfaces. The lattice type and NP spacing can be readily tuned by adjusting structural parameters of HINPs, such as polymer length and the relative size of NP cores.[52] Recently, Xu *et al.* reported the self-assembly of PS-tethered AuNPs and Fe_3O_4 NPs into a variety of complex 2D and 3D binary nanocrystal superlattices (BNSLs) in diethylene glycol/air interface (Figs. 4.5(d–f)). The formation of different BNSLs was controlled by tuning the size of NP core and MW of densely grafted PS.[45] This drying-mediated assembly provides a route to engineer the interactions and orientations of NPs into densely packed arrays, e.g., within 2D interface.

4.4.1.2. *Assembly within BCP film*

Polymer matrices (e.g., BCPs) can be used to induce the anisotropic self-assembly of HINPs into flexible nanocomposites.[53,54] The

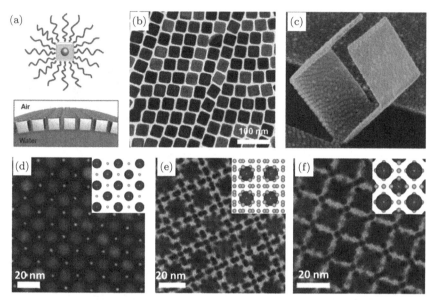

Fig. 4.5. Air-liquid interface mediated self-assembly of HINPs. (a) Schematic of a hairy AgNC and the assembly of the cubes at the air-water interface. (b, c) Images of assembled plasmonic nanosheet (b) and corresponding origami structure fabricated from the nanosheet (c). (d–f) Representative structures of BNSLs self-assembled from PS-tethered 3.8 nm AuNPs and 13.4 nm Fe_3O_4 NPs: AB-type 2D BNSLs (d), $NaZn_{13}$-type (e) and bcc-AB_6-type 3D BNSLs (f). Insets are corresponding structural model. Reprinted with permission from Refs. 45 and 51. (Copyright 2014 Material Research Society and 2015 Nature.)

wettability of the HINPs in surrounding matrix plays a crucial role in determining the segregation or dispersion, orientation, and spatial arrangement of HINPs within the matrix.[27,55,56] The wettability of HINPs can be tailored by the chemical composition, architecture, and σ of polymer ligands.[57] The interactions between brushes and matrices can be summarized into three typical categories:

(1) Favorable brush−matrix interactions. It is beneficial to the uniform dispersion of HINPs within matrices. This can be generally achieved by selecting polymer brushes that have attractive interactions with the matrix (i.e., a Flory−Huggins interaction parameter, χ that is less than zero).

(2) Incompatible interactions between brushes and matrices. The phase-separation occurs when HINPs are embedded in incompatible matrices. In this case, the spatial arrangement of HINPs

is largely governed by the interparticle interactions. For example, Tao *et al.* reported two typical self-orientation models (i.e., face-*to*-face and edge-*to*-edge) of AgNCs tethered with hydrophilic thiol-terminated PEG or poly(vinyl pyrrolidone) (PVP) within hydrophobic PS matrices (Fig. 4.6).[55] The NPs organized into linear chains in the polymer matrix, after the solvent or thermal annealing of polymer nanocomposites.[55,58] The orientation

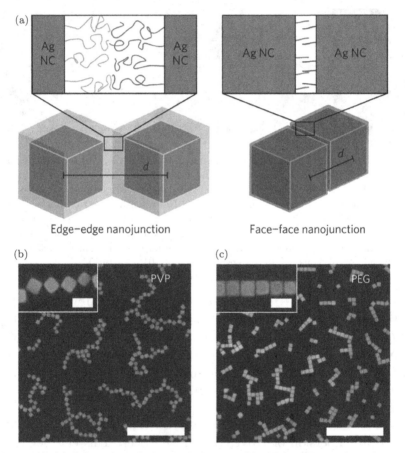

Fig. 4.6. (a) Schematics illustrating the assembly of AgNCs in edge–edge and face–face orientation modes by grafting nanocubes with polymers of different lengths. (b, c) SEM images of linear chains of oriented polymer-grafted AgNCs assembled in PS film. Scale bar in large and inset images are 1 μm and 100 nm, respectively. AgNCs are modified with (b) PVP (MW of 55 Kg/mol) and (c) PEG (4 repeating units of PEO) ligands. Reprinted with permission from Ref. 55.

of NPs is dictated by the steric interactions between polymer brushes from neighboring NPs.

(3) Compatible interactions between brushes and matrices. In this case, the polymers on NPs and as matrices (or one of the matrix components) possess identical chemistry. When dispersed in compatible matrices, self-organization of HINPs can be influenced by conformational entropy arising from the compression and extension of brushes.[24,59] At relatively high grafting densities (in SDPB/or CPB regime) of polymer ligands, HINPs preferentially segregate to reduce the entropic penalty caused by the stretching of polymer brushes. Such significant depletion attraction is known as autophobic dewetting effect,[60] which decreases with increasing NP surface curvature, that is, small spherical NPs possess less entropic stretching penalty than large ones or rod-like NPs. For example, 3.6 nm cadmium sulfide coated with PS_{120} brushes (PS-CdS) could be well dispersed and incorporated in PS phase of PS_{1200}-b-PEO_{370} matrices, after the co-assembly at the air-water interface (Fig. 4.7). Such small CdS cores make PS-CdS wetted by PS-b-PEO matrices, while the *di*-BCPs formed strips. The relative high σ of PS chains on CdS NPs allowed for the PS shell to shield the attraction interactions between CdS cores.

Conversely, HINPs with a low σ of polymers may experience anisotropic interactions when incorporated within the polymer matrix. For example, SiO_2 NPs with sparsely grafted PS can assemble into strings, sheets, and platelet-like structures in a PS matrix, as a result of phase-separation between the matrix and exposed surface (i.e., ungrafted with polymers) of NPs.[19]

Assemblies of supramolecules can be used to guide the organization of HINPs into ordered architectures. One typical system is constructed by hydrogen bonding of 3-pentadecylphenol (PDP) with pyridine groups of polystyrene-b-poly(4-vinylpyridine) (PS-b-P4VP) (PS-b-P4VP(PDP)) (Fig. 4.8).[62,63] For example, Zhu *et al.* reported the uniform dispersion of PS-coated AuNPs within cylindrical PS phases of PS-b-P4VP(PDP) assemblies. The linear NP arrays can be released from the films by breaking down hydrogen bonds in ethanol

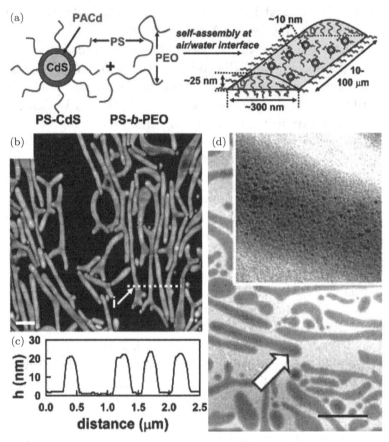

Fig. 4.7. (a) Schematic illustrating the co-assembly of PS-coated cadmium sulfide NPs (PS-CdS) and PS-*b*-PEO at the air—water interface. Representative (b) AFM and (d) TEM images of the resulting hierarchical assemblies. (c) The height of the structures corresponding to the dashed line in (b). Reprinted with permission from Ref. 61.

(Fig. 4.8(a)).[62] Nie and Zhu *et al.* also showed that the assembly morphology transitioned from spheres, to cylinders, to nanosheets with increased loading of PS-tethered AuNPs or decreased ratio of PDP to 4VP.[64] When PS-tethered AuNRs were used, strong depletion attractions arising from autophobic dewetting drove the hairy AuNRs to assemble side-by-side to form an ordered smectic B phase.[65] It was found that the grafting of binary PS brushes with different lengths can largely suppress the autophobic dewetting effect,[66]

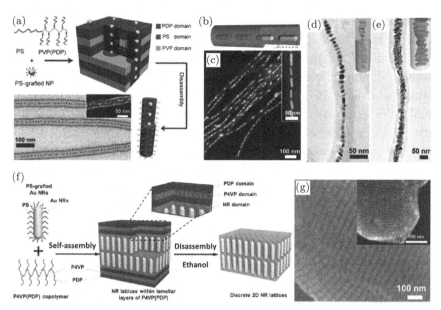

Fig. 4.8. Supramolecular BCP-guided self-assembly of HINPs. (a) Schematic illustrating the assembly of PS-tethered AuNPs in cylindrical PS domains assembled from comb–coil supramolecules of PS-*b*-P4VP(PDP). (b) TEM images o of PS core of cylindrical micelles encapsulated with organized hairy AuNRs grafted with binary polymer brushes: (c) PS_{51k}-*b*-$P4VP_{18k}(PDP)_{2.0}$, NR with $d \times L$: 7 nm × 29 nm; (d) PS_{51k}-*b*-$P4VP_{18k}(PDP)_{2.0}$, NR of 6 nm × 12 nm; (d) PS_{110k}-*b*-$P4VP_{107k}$ $(PDP)_{1.0}$, NR of 7 nm × 29 nm. (f) Schematic illustrating the assembly of PS-grafted Au NRs into superlattices mediated by supramolecules of $P4VP(PDP)_{1.0}$. (g) Representative SEM images of side and top (inset) view of 3D NRs superlattices. Reprinted with permission from Refs. 62, 65 and 67. Copyright 2011 and 2015 Wiley-VCH, 2013 American Chemical Society.

thus significantly improving the wettability and uniform dispersion of NRs within PS domains. The orientation of AuNRs can be controlled to be parallel or perpendicular to cylindrical PS domains (Figs. 4.8(b–d)). Long AuNRs preferentially aligned parallel to PS domains, in order to minimize the entropic penalty associated with the deformation of the surrounding PS matrix (Fig. 4.8(b)). In contrast, short rods aligned perpendicular to PS domains at a relatively high volume fraction of NRs (Fig. 4.8(c)). When the diameter of cylindrical micelles was large, long NRs hexagonally packed within the cylinders and twisted along the cylindrical axis to relief the above-mentioned entropic penalty and excluded volume of polymer-capped ends of the NRs (Fig. 4.8(d)). More recently, Nie and Zhu *et al.* utilized lamellar structures of P4VP (PDP) assemblies as templates to organize PS-capped AuNPs with different shapes (e.g., spheres, rods, and cubes) into free-standing 2D and 3D superlattices with controllable spacing (Figs. 4.8(e) and (f)).[67] It was also found that PDP molecules can serve as plasticizers that drastically enhance the mobility of HINPs in the assembly process, thus enabling the efficient packing and orientation of NPs to form superlattices. This work suggests that the enthalpic and entropic interactions between HINP brushes and matrix can be modulated by controlling the σ of polymer hairs and hence the wettability of HINPs. Thus, wettability control provides an effective route to direct the organization of HINPs, in addition to the templating effect of microphase-separation of BCP matrix.

Furthermore, co-assembly of HINPs and BCPs within 3D confined space (e.g., emulsion droplet) have been proved to be a promising strategy for preparing functional hybrid materials with more complexity and functionality.[68–70] For example, Jiang *et al.* demonstrated that both the location and alignment of AuNPs on scaffolds of PS-*b*-P4VP assemblies can be controlled through *co*-assembly in 3D confined space.[68] A solution of mixed hairy AuNPs and BCPs were first emulsified into oil-in-water emulsions in the presence of surfactants. The subsequent evaporation of solvents in the emulsions triggered the migration and organization of AuNPs relative to the BCPs and oil/water interface, thus resulting in the formation of BCP colloidal assemblies with precisely positioned AuNPs at specific locations. The migration and organization of hairy AuNPs are attributed to the entropic effect arising from the conformational entropy loss of the polymer chains, as well as the enthalpic attraction between brushes on AuNPs and surfactants at the oil/water interface.

4.4.1.3. *Assembly of neat HINPs*

In addition to the use of BCP templates, anisotropic self-assembly of neat HINPs has emerged as an attractive strategy to fabricate hybrid composites that may exhibit unique properties.[12,57,71,72] As an example, Vaia and co-workers reported the assembly of PS-grafted silica NPs into nanocomposite materials with a nonisotropic local arrangement of NPs.[72] The organization of NPs within the composites is strongly dependent on the σ of PS on the NPs. It is remarkable that the composites exhibit viscoelastic behavior during elongation, resembling that of semi-crystalline elastomers.

Moreover, phase separation of immiscible polymer brushes on individual HINPs can also drive the anisotropic assembly of neat HINPs into ordered morphology in nanocomposite films. For instance, Matsushita and *co*-workers reported that AuNPs grafted with a binary mixture of immiscible polyisoprene (PI) and PS assembled into lamellar structures with NPs located at the PI–PS interface.[57] The corresponding 2D SAXS patterns and TEM images showed the anisotropy of the nanocomposite film structures from different observing directions (Fig. 4.9). The random arrangement of Au NPs located within the PI-PS interface indicates that the interactions between NP cores are too weak and totally overcome by the phase-separation of binary polymer brushes (Fig. 4.9(d)).

Despite the initial success in the assembly of neat HINPs, relatively small HINPs (usually with the size of the inorganic core $<10\,\mathrm{nm}$) are often used in current studies. This is largely due to the low mobility of larger HINPs and the difficulty in achieving equilibrium of assembled structures. There is an urgent need of new assembly strategies in the field, in order to expand the size range of HINPs that can be assembled.

4.4.2. *Self-assembly of HINPs in solutions*

The spontaneous generation of surface anisotropy on isotropically functionalized NPs makes the assembly of HINPs particularly intriguing, as it does not require sophisticated asymmetric surface functionalization. In this case, directional interactions arise from the conformational change of polymer brushes and/or the phase separation of multiple immiscible polymer brushes on NP surfaces during the assembly of HINPs in solutions.[21,57,73]

4.4.2.1. *HINPs tethered with mixed polymers*

HINPs isotropically decorated with a mixture of homopolymers can readily develop surface anisotropy due to the lateral phase-separation

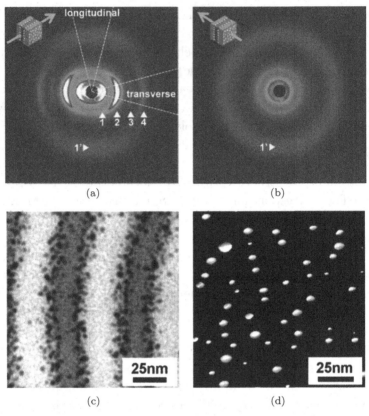

Fig. 4.9. Spatial arrangement of neat HINPs within nanocomposites. 2D SAXS patterns from edge-view showing an anisotropic pattern (a) and from through-view shows an isotropic pattern (b), respectively. (c) TEM micrograph of solvent-annealed AuNP-IS hybrid film. The observing direction is edge-view. Since the sample was stained with OsO_4, PI and PS phase appear dark and bright, respectively. AuNPs appear black dots in c due to the highest electron density among the components. (d) Reconstructed TEMT image, where the white dots correspond to AuNPs. The observing direction is through-view. The obtianed hybrid nanocomposite film with the volume fraction of polyetyrene $\varphi_S = 0.48$. (Reprinted with permission from Ref. 57.)

of immiscible polymers.[21,34,73,74] Such induced anisotropies of HINPs with amphiphilic binary brushes could direct the self-assembly of nanocomposites by themselves in selective solvents. A pioneered work by Zubarev *et al.* reported the self-assembly of 2 nm AuNPs chemically linked with V-shaped $PEO_{50}-b-PS_{40}$ into rods, wormlike micelles, and vesicles with arrays of NPs localizing at the interface of

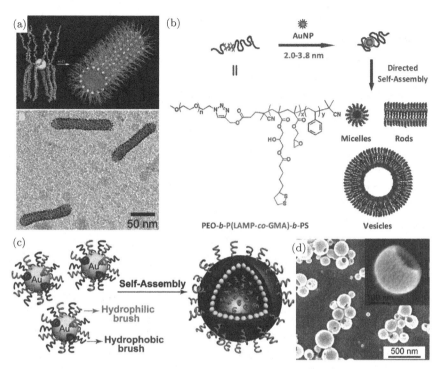

Fig. 4.10. Self-assembly of HINPs with binary immiscible brushes. (a) Illustration of the self-assembly of V-shaped polymer-modified AuNPs and TEM image of corresponding assemblies. (b) Schematic illustrating the fabrication of ultrasmall AuNPs (2.0∼3.8 nm) anchored with a single tri-BCP and their self-assembly in selective solvent. (c) Schematic illustration of the self-assembly of amphiphilic NPs with mixed polymer brushes into vesicular structures. (d) SEM images of the plasmonic vesicles assembled from 14 nm AuNP grafted with mixed PEO/PMMA brushes. Reprinted with permission from Refs. 32, 40 and 41. Copyright 2006, 2012 and 2011 American Chemical Society.

PS/PEO (Fig. 4.10(a)).[40] For vesicles, bilayer of NPs were arranged at the inner and outer interfaces. Similarly, Moffitt *et al.*[28] constructed responsive HINPs of CdS QDs coated with tri-BCPs of polystyrene-block-poly(acrylic acid)-block-poly(methyl acrylic acid) PS-PAA-PMAA. During the assembly, the PS and PMAA blocks acted as mixed brushes with the middle PAA block cross-linked to lock the CdS inside. Similar ensembles were also obtained by Liu *et al.* using monofunctinalized 2.0–3.8 nm AuNPs modified with a single *tri*-BCP chain consisting of PEO and PS outer blocks and a 1,2-dithiolane-functionalized Au-binding middle block (Fig. 4.10(c)).[41]

Such HINPs resemble amphiphilic *di*-BCPs to assemble into different structures. The assembly process can be attributed to the wrapping of brushes initially isotropically distributed on NP surface to form a Janus surface and the subsequent segregation of the hydrophobic PS phase. Small inorganic NPs ($d <$ ca. 8 nm) within such HINPs often fluctuate at the interface of binary immiscible brushes to form random arrays within the assemblies,[57] probably due to the relatively weak interparticle interactions that are shielded by the relatively crowd brushes. The morphological transition can be readily tuned via the hydrophilic/hydrophobic balance of mixed brushes.

Larger NPs may be favorable for many of their potential applications due to the size dependent properties or coupling interactions between these NPs. It is believed that the anisotropic self-assembly of HINPs with relatively broad size range (e.g., being large in diameter) inorganic cores would be beneficial to the property discovery and application development of NPs. Song *et al.* reported the assembly of relatively large AuNPs (~14 nm nanospheres, or 13×46 nm NRs) grafted with binary polymer brushes into plasmonic gold vesicles (GVs) with a monolayer of AuNPs in the membrane (Figs. 4.6(c) and (d)).[32,33] The self-assembly was triggered by film rehydration method widely used for molecular self-assembly.[75] The obtained vesicular morphology with single layer of AuNPs indicates that the isotropically grafted polymer brushes on HINPs rearranged to form a patchy topology when exposed to selective solvents, followed by aggregation of the hydrophobic patches (PMMA or PMMAVP) to form the vesicle walls. Different combinations of polymers such as PEO/poly(methyl methacrylate-*co*-4-vinyl pyridine) (PMMAVP), PEO/PMMA, and PEO/polylactide (PLA) have been used to introduce desired responsive functionality. For example, PMMAVP-based vesicular assemblies are pH-responsive; the vesicles can be dissociated to release payloads at an acid environment (pH = 5).[76]

4.4.2.2. *HINPs grafted with amphiphilic BCPs*

It is often believed that a mixture of hydrophilic and hydrophobic species extended directly from NP surface is required to induce amphiphilicity of NPs. This concept is routinely adopted for NP self-assembly including the "V-shaped BCP" system. Interestingly, Nie *et al.* recently showed that the conformation change of linear BCP tethers (or more complex multi-block copolymers) can lead to controllable assembly of HINPs into various nanostructures. This method may offer the following advantages over existing approaches: i) reduced constraints of bulky NP cores on the conformational flexibility of polymer chains; ii) precisely controlled chemical composition and functionality on NP surface; and iii) richer complexity in the architecture of multi-block copolymers. Given current advances in the synthesis of BCPs with excellent controllability and complexity, this general approach could open up new realms of possibilities to create libraries of novel colloidal building blocks with predesigned surface chemistries.

4.4.2.2.1. Self-assembly by film rehydration

Nie *et al.* demonstrated that HINPs consisting of AuNPs grafted with thiol-terminated PS-*b*-PEO can self-assemble into spherical hollow vesicles and tubular structures via film rehydration (Fig. 4.11).[42] Both nanostructures were composed of a monolayer of hexagonally packed AuNPs in the membrane (Fig. 4.11(b)). The assembly morphologies are predominantly dependent on the hydrophilicity of the HINPs, which is determined by the ratio between the average root-mean-square end-to-end distance (R_0) of the PS block and the size of AuNP (d_{Au}). As shown in a phase-like diagram in Fig. 4.11(e), HINPs favor vesicle formation when $R_0/d_{\mathrm{Au}} < 0.5$, tubule formation when $R_0/d_{\mathrm{Au}} \approx 0.5$, and cannot be rehydrated when $R_0/d_{\mathrm{Au}} > 0.5$. The morphological transition is attributed to the surface curvature-dependent hydrophilicity of the HINPs, which is determined by the density of hydrophilic polymer tips on NP surfaces (ρ, defined as $\rho = \sigma/[4(0.5 + R_0/d_{\mathrm{Au}})^2]$. When $R_0/d_{\mathrm{Au}} > 0.5$, a low ρ makes the HINPs too hydrophobic to be rehydrated and assemble in water. It is worth noting that a higher σ of polymer tethers can compensate the reduction of hydrophilicity at elevated R_0/d_{Au}, thus triggering the assembly of HINPs.

Other than morphological control, the length of PS also determines the interparticle spacing, D_{Au}, between adjacent NPs.

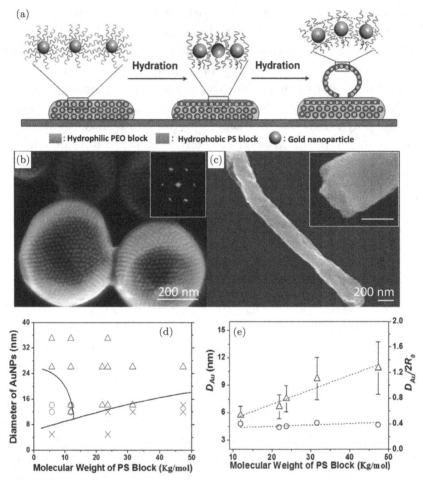

Fig. 4.11. (a) Schematic illustrating the self-assembly of AuNPs grafted with PS-*b*-PEO via film rehydration. (b, c) representative SEM images of assembled vesicles and tubules. Inset in (b) is the Fast Fourier Transform (FFT) pattern of the SEM image. (d) Product diagram for the assembly of HINPs with varying diameter of Au NPs and MW of the PS block: $\triangle$, vesicles; $\circ$, tubules; and $\times$, precipitates. (e) Plot of interparticle spacing D_{Au}($\triangle$) and $D_{Au}/2R_0$ ($\circ$) as a function of the MW of the PS block for vesicular structures. (Reprinted with permission from Ref. 42. Copyright 2012 American Chemical Society.)

Figure 4.7(e) shows that the average D_{Au} of such vesicles approximately linearly increased from 5.7 ± 0.9 to 10.9 ± 2.9 nm with increasing PS MW from 11.9 to 47.3 Kg/mol. Such precise control over D_{Au} allows for tuning of the coupling interactions between NPs and thus their plasmonic properties. The D_{Au} was found to be reduced from 19.3 to 7.89 nm after the formation of vesicles. This suggests that segments of grafted polymers may relocate from the gap between pairs of NPs towards both sides of NPs (or vesicular membranes). As a result, the spontaneous generation of particle anisotropy arises from conformational changes of BCP brushes, in order to minimize the interfacial energy by maximizing the exposure of the hydrophilic blocks to water, while shielding the hydrophobic blocks.

4.4.2.2.2. Self-assembly in selective solvent

Nie *et al.* systematically explored the role of kinetic and thermodynamic factors in the self-assembly of HINPs into various nanostructures in selective solvents (THF/H$_2$O).[77] During the self-assembly process, the hydrophobic PS blocks collapse to minimize the overall free energy of the system, leading to the formation of various nanostructures. The dimensions and morphologies of the assemblies are strongly dependent on NP size and PS length. The morphology changes from hollow vesicles, to small clusters, and to unimolecular micelles, as the PS length decreases relative to the AuNP size.

Dissipative particle dynamics (DPD) simulations with coarse-grained potentials indicate that the reorganization of brushes on NP surfaces and the deformability of the polymer shell play a crucial role in the morphological transition of the assemblies. When the flexible PS blocks are too short relative to the AuNP size, the PS blocks cannot stretch and compact sufficiently to generate an anisotropic surface; thus, the HINPs preferentially aggregate into small clusters or unimolecular micelles. Once the PS block is sufficiently long, the HINPs can be effectively deformed, due to the redistribution of the flexible long polymer tethers. The resulting HINPs behave as colloidal analogies of ABA *tri*-BCPs consisting of a middle hydrophobic section and two hydrophilic side sections, to assemble into various architectures.

The σ of BCP brushes on NPs plays a crucial role in determining the directional interactions between HINPs and their assembly paths into different architectures. We demonstrated that depending on σ, PS-*b*-PEO tethered AuNPs can undergo step-wise self-assembly into

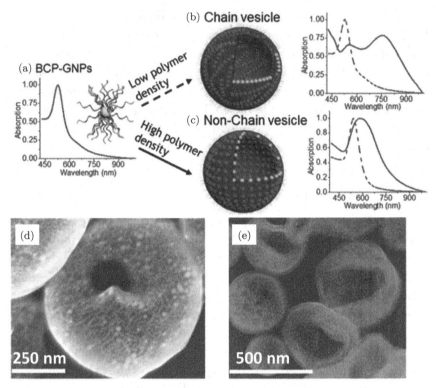

Fig. 4.12. (a) Schematic illustrating the σ-dependent self-assembly of PS-*b*-PEO grafted AuNPs into (a) chain vesicles (b) and non-chain vesicles (c). (d,e) SEM images of chain vesicle (d) and non-chain vesicle (e). (Reprinted with permission from Ref. 46. Copyright 2015 Wiley-VCH.)

chain vesicles composed of linear NP strings in the membrane: the HINPs assembled into NP strings that subsequently folded up to form chain vesicles (Fig. 4.12).[46] At a low $\sigma = \sim 0.03$ chain/nm^2, HINPs made from 13 nm AuNPs self-assemble into chain vesicles (Fig. 4.12(d)). At a high $\sigma = \sim 0.08$ chain/nm^2, nonchain vesicles with uniform distribution of NPs in the membrane were formed (Fig. 4.12(e)). The interparticle distance of chain vesicles is ~ 0.8 nm, which is much smaller than a value of 9.0 nm for nonchain vesicles. As a result, chain vesicles exhibit a significant red-shift in the localized surface plasmon resonance (LSPR) peak from 540 to ~ 760 nm due to a strong plasmonic coupling between NPs, while the peak was only shifted to 580 nm for nonchain vesicles.

The σ-dependent assembly of HINPs constitutes a novel assembly mechanism. In the initial stage, the formation of NP strings is a consequence of many-body interactions between HINPs.[19] When two HINPs with a low σ interact due to solvophobic attraction, the redistribution and deformation of polymer hairs creates a high-density polymer region near the point of contact. Consequently, a third particle preferentially approaches from the ends due to the attractive van der Waals forces, rather than sides of the cluster due to increased steric hindrance of the high-density hairs. This is evidenced by the larger separation distance between strings ($\sim$12 nm) in vesicular membranes than spacing between neighboring AuNPs in each string ($\sim$0.8 nm). By contrast, interactions among HINPs with high σ are mediated by solvophobic forces rather than van der Waals forces. These many-body interactions lead to 2D planar films that better shield the hydrophobic PS from solvent media, thus eventually the formation of non-chain vesicles. This work demonstrates the possibility of achieving high complexity and hierarchy in HINP assembly via engineering the directional interactions between HINPs by taking advantage of the conformational changes of long flexible polymers (i.e., longer brushes relative to NP core and low σ).

4.4.2.2.3. Concurrent self-assembly of amphiphilic HINPs with BCPs

Concurrent self-assembly of multiple building blocks has emerged as a powerful strategy for preparing nanostructures with more complex architectures and richer functionalities.[78] Nie *et al.* showed that amphiphilic HINPs composed of BCP-tethered AuNPs can co-assemble with free amphiphilic BCPs of PS-*b*-PEO to generate a variety of hybrid vesicles with well-defined shapes, morphologies, and surface patterns in selective solvents (Fig. 4.13).[79] They include patchy vesicles (PyVs) with multiple small HINP domains, Janus-like vesicles (JVs) with two distinct halves, and heterogeneous vesicles (HVs) with uniform distribution of HINPs. In particular, the interplay of bending modulus and line tension between BCP and HINP domains gives rise to hybrid JVs with

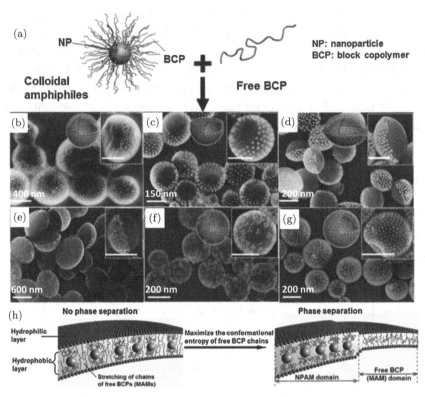

Fig. 4.13. (a) Schematic illustrating the co-assembly of a mixture of amphiphilic PS-*b*-PEO grafted AuNPs and free BCP PS-*b*-PEO. (b–e) Representative SEM images of Janus vesicles with (b) spherical, (c) hemispherical, and (d, e) disk-like shapes. SEM images of (f) patchy vesicles and (g) heterogeneous vesicles. (h) Schematic illustration of the lateral phase separation mechanism of building blocks in vesicular membranes. (Reprinted with permission from Ref. 79. Copyright 2014 American Chemical Society.)

intriguing shapes, including spherical, hemispherical, and disk-like shapes (Figs. 4.13(b–e)). Compared with BCP domains, the Young's modulus of HINP domains can be readily tuned in a larger range by controlling the relative dimensions of the soft polymer shell and the rigid inorganic core. DPD simulations suggest that the lateral phase-separation of BCPs and HINPs in vesicular membranes is driven by the conformational entropy gain of flexible polymer tethers, arising from the mismatch in the effective membrane thickness of BCP and HINP domains (Fig. 4.13(h)). Morphological transitions of assemblies from PyVs, to JVs, and to HVs can be tuned by varying the PS length of brushes.

The co-assembly strategy can be extended for fabricating composite vesicles containing multiple building blocks, in order to further enrich the functionality and complexity of assembled structures. For example, magneto-plasmonic Janus vesicles (JVs) asymmetrically integrated with AuNPs and magnetic NPs (MNPs in the membrane) can be prepared by co-assembling PS-*b*-PEO grafted AuNPs, free PS-b-PAA and hydrophobic MNPs (Fig. 4.14).[80] Figure 4.14(d) shows

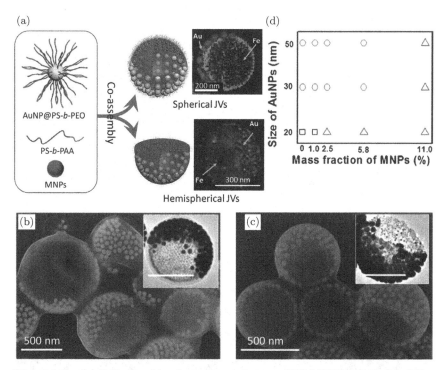

Fig. 4.14. (a) Self-assembly of a ternary mixture of PS-*b*-PEO tethered AuNPs, MNPs of Fe_3O_4 NPs, and free BCPs of PS-*b*-PAA into hybrid magneto-plasmonic JVs with spherical and hemispherical morphologies. EDS mapping of Fe (green) and Au (red) further confirms the structures. (b, c) SEM and TEM (inset) images of the corresponding JVs. (d) Phase-like diagram for the formation of hybrid vesicles with different morphologies by varying the core size of BCP-tethered AuNPs and mass fraction of MNPs: spherical heterogeneous vesicles (square), spherical JVs (circle), and hemispherical JVs (triangle). (Reprinted with permission from Ref. 80.)

that the morphological transition of the magneto-plasmonic vesicles is strongly dependent on both the core size of PS-*b*-PEO grafted AuNPs and the mass fraction of hydrophobic Fe_3O_4 NPs. The formation of Janus vesicles of different shapes (spherical vs. hemispherical) is attributed to the mismatch in the Young's modulus of polymer and HINPs domains.

In another interesting work, magneto-plasmoic vesicles were assembled from Janus Au-Fe_3O_4 NPs surface modified with one of binary homopolymer ligands (Fig. 4.15).[15] The Au-Fe_3O_4 vesicles were distinguished by the vesicular structures with double layered

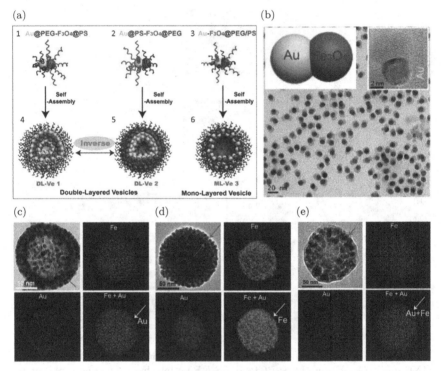

Fig. 4.15. Self-assembly of Janus NPs regioselectively grafted with binary polymers. (a) Schematic illustration of the Janus Au-Fe_3O_4 NPs grafted with hydrophilic PEG on Au and hydrophobic PS on Fe_3O_4 (1), with PS on Au and PEG on Fe_3O_4 (2), and with binary mixed PEG and PS (3), and their self-assembly into double-layered plasmonic-magnetic vesicles and mono-layered vesicles in aqueous media. (b) TEM image of the Janus Au-Fe_3O_4 NPs. (c–e) TEM-element mapping images of the DL-Ve 1 (c), DL-Ve 2 (d), and monolayered vesicles (e). (Reprinted with permission from Ref. 15.)

inorganic NPs. Regiospecific surface modifications of Janus NPs make the HINPs resemble amphiphilic '*di*-BCPs', which determines the organization and location of Janus NPs in the vesicular membranes. TEM mapping confirms the formation of different structures, namely, mono-layered vesicles, double-layered vesicles with Fe_3O_4 NPs in the membrane center (DL-Ve1), and double-layered vesicles with AuNPs in the membrane center (DL-Ve2). Thanks to the interparticle plasmonic coupling of AuNPs and magnetic diploe interaction of Fe_3O_4 NP, the double-layered vesicles exhibited greatly enhanced optical and magnetic properties that facilitate their biological performance.

4.4.2.2.4. Kinetically-controlled self-assembly in microfluidics

As mentioned above, the mobility and relaxation of HINPs is significantly slower than molecules, suggesting a critical role of kinetics in HINP self-assembly. Microfluidics offers a tool to control the kinetics of HINP assembly to produce various architectures.[81,82] In this method, the HINP assembly is conducted by forcing a solution of HINPs in THF between two water streams through a joint downstream channel to form a laminar flow in flow focusing microfluidic devices (MFFDs). The diffusive mixing between laminar fluids creates a well-defined solvent gradient to trigger assembly of HINPs. By varying flow hydrodynamics, the assembly process generates various non-equilibrium structures, such as micelles, giant vesicles, and disks in MFFDs (Fig. 4.16). Take PS-*b*-PEO tethered AuNRs as an example, the increase in the volumetric flow rate ratio of THF to water (Q_{THF}/Q_{H2O}) led to the transition of assembled structures from micelles with diameters less than 100 nm to giant vesicles with diameters greater than 500 nm, and to disk-like monolayers of AuNRs. When $Q_{THF} \ll Q_{H2O}$, small micelles were produced due to the fast depletion of the HINP phase and quenching of the assembly process. At a higher Q_{THF}/Q_{H2O}, the slower mixing between streams allowed for the formation of giant vesicles. Further increase of Q_{THF}/Q_{H2O} led to the formation of disk-like micelles.

The size of HINP assemblies can be controlled by varying Q_{THF}/Q_{H2O}.[82] Generally, the size of assembled vesicles increased with increase in the value of Q_{THF}/Q_{H2O}. In addition, the transverse

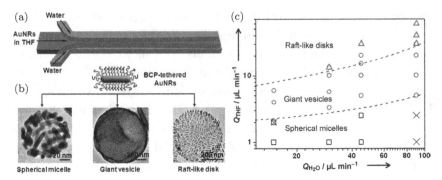

Fig. 4.16. (a) Hydrodynamic self-assembly of amphiphilic NPs tethered with BCPs using microfluidic flow-focusing devices (MFFDs). A THF solution of polymer-tethered NPs is focused by two water streams on the sides. (b) TEM images of assemblies of PEO_{45}-b-PS_{211} tethered AuNRs at different hydrodynamic conditions. (c) The product diagram of the self-assembly of polymer-tethered AuNRs in the $[logQ_{THF}, logQ_{H2O}]$ space: micelles (square), disks (up triangle), giant vesicles (circle), and no assemblies (cross). (Reprinted with permission from Ref. 83. Copyright 2013 Wiley-VCH.)

diffusion of HINPs across the lamellar boundary is also critical for their assembly. Under the same flow conditions, larger HINPs produced larger vesicles due to the reduced diffusion rate. Moreover, this microfluidic assembly can be extended to fabricate hybrid nanostructures with more complex morphologies (e.g., hybrid JVs) by co-assembly of a mixture of multiple building blocks (e.g., PS-b-PEO tethered HINPs and free PS-b-PEO).[81] The overall size of JVs can be controlled in the range of 400–2600 nm and the surface area of each half of the hybrid JVs can be varied by changing the weight ratio of building blocks.

4.5. Biomedical Applications of Vesicular Nanocomposites

Among different nanostructures, organic vesicles (e.g., polymersomes and liposomes) have made great impact as promising nanocarriers for bioimaging and drug delivery. Compared with organic vesicles, hybrid vesicles integrated with inorganic NPs such as AuNPs and/or MNPs in the membranes possess superior optical, electronic, or magnetic properties originated from the inorganic components. These vesicles show important features, such as high loading efficiency of therapeutic agents, enhanced imaging capability, externally controlled release, etc. Representative systems such as plasmonic GVs and magnetic vesicles (MVs) have shown reasonable potential for cancer imaging, cancer therapy, and imaging-guided drug delivery.[22,33,43,46,76,77,84,85]

The hybrid vesicles can be designed to achieve externally controlled release of loaded therapeutic agents by utilizing the responsiveness of polymer ligands, hyperthermia effect of integrated inorganic NPs, or the synergetic interactions between polymers and NPs. One typical design of, the disease-specific release of payload is to introduce stimuli-responsive hydrophobic-to-hydrophilic transition or degradation to polymer ligands and hence the dissociation of vesicles by chemically tailing the hydrophobic segments of polymers. The release and pharmacokinetics of payload can often be tracked or assessed in real time by monitoring the variation in imaging signal through different imaging modalities, such as fluorescence, surface-enhanced Raman scattering (SERS), and magnetic resonance imaging (MRI) enabled by the hybrid vesicles.[22] For example, Duan *et al.* explored the application of pH-sensitive GVs assembled from AuNPs grafted with binary PEG/PMMAVP ligands in the biological environment.[33] The attachment of a HER2 antibody at the ends of PEO brushes enabled the specific recognition of the GVs to SKBR-3 breast cancer cells with overexpressed HER protein. As illustrated in Fig. 4.17(a), once the GVs were taken up by the cells, the acidic tumor environment caused the protonation of pyridine groups and hence the hydrophobic (25% of vinylpyridine) hydrophilic transition of PMMAVP, resulting in the disruption of vesicular membranes. The associated changes in optical signals (i.e., blue shifts in scattered light and decrease in SERS intensity) were employed to trace the intracellular release of a model drug, doxorubicin (DOX) (Figs. 4.17(b, c)). The intracellular redistribution of DOX was found to be quantitatively correlated to the feedback of scattering light and SERS signals.

The organization of NPs within vesicular membranes offers an effective route to control the coupling interactions (e.g., plasmonic coupling and magnetic-magnetic interactions) between NPs, thus facilitating the biological performance of the hybrid vesicles. Take GVs as an example, the assembly of AuNPs allows for tuning the

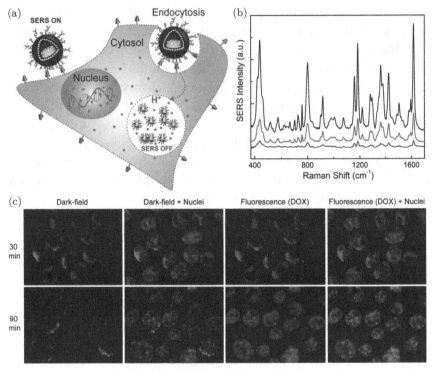

Fig. 4.17. (a) Schematic illustration of the cellular binding and pH-regulated intracellular drug release of the SERS-encoded pH-sensitive GVs assembled from PEG/PMMAVP grafted AuNPs. (b) SERS spectra of SKBR-3 cells treated with GVs after 30 min incubation (black line) and the post incubation spectra of the cells at 60 min (red line) and 90 min (blue line). (c) Dual dark-field and fluorescence imaging of SKBR-3 cells labeled with doxorubicin (DOX) loaded GVs after 30 and 90 min incubation. (Reprinted with permission from Ref. 33. Copyright 2012 American Chemical Society.)

localized surface plasmon resonance (LSPR) of the assemblies to near-infrared wavelength range by varying the MW of polymer ligands and the interparticle spacing of NPs. The vesicles with strong NIR absorption are more ideal for *in vivo* applications than spherical AuNPs which show strong absorption in the visible range, due to the deeper penetration of NIR light into tissues than visible light. The strong NIR absorption also makes them promising photothermal contrast agents that can efficiently turn absorbed light to heat upon NIR irradiation, thus enabling their use in effective cancer imaging and therapy. Nie *et al.* demonstrated the utilization of GVs assembled from 26 nm PS-*b*-PEO tethered AuNPs for *in vitro* and *in vivo* photothermal and photoacoustic imaging of tumors (Figs. 4.18(a–c)). After intratumoral injection of GVs into mice with subcutaneous MDA-MB-435 breast cancer, a 3.8-times increase of photoacoustic signal in the tumor region was observed (Fig. 4.18(c).[85] The imaging performance of GVs can be further improved by reducing the interparticle spacing between NPs in the membranes, as reflected in an exemplary biodegradable GVs (BGVs)

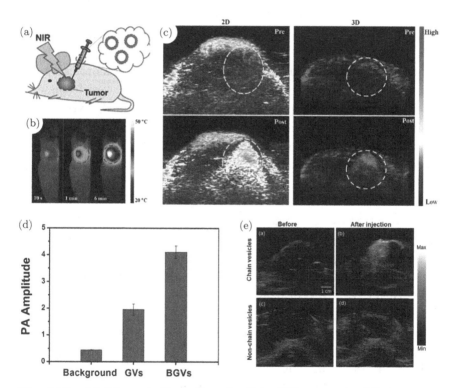

Fig. 4.18. (a) Schematic illustration of utilizing NIR-responsive GVs for cancer imaging. (b) Photothermal imaging of tumors in mice after different irradiation times. (c) 2D and 3D photoacoustic imaging of tumors in mice under laser irradiation before and after injection of GVs. (d) A comparison of *in vivo* photoacoustic signal for GV and BGV system. (e) Photoacoustic imaging of mouse tissue before and after injection of chain vesicles or non-chain vesicles. NIR laser irradiation was used in all the cases. (Reprinted with permission from Refs. 43, 46, 85. Copyright 2013 American Chemical Society, 2013 and 2015 Wiley-VCH.)

assembled from 26 nm poly(ethylene glycol)-*b*-poly(caprolactone) (PEG-*b*-PCL) grafted AuNPs.[43] Preliminary *in vivo* studies showed that photoacoustic signal of BGVs was doubled, compared with GVs (Fig. 4.18(d)). The organization of NPs in the vesicular membranes also plays a crucial role in determining the optical response and hence their imaging performance of the hybrid vesicles. When AuNPs are arranged into chains in the membrane, the NIR absorption of resulting chain vesicles is much higher than non-chain vesicles (i.e., vesicles with uniform distribution of NPs in the membranes), due to a close interparticle distance ($\sim$0.8 nm) in the AuNP strings.[46] In *in vivo* studies, an about 7–8 times increase in photoacoustic signal was observed for chain vesicles compared with non-chain vesicles (Fig. 4.18(e)).

Other than spherical NPs, shaped NPs such as nanorods can be also used to construct hybrid vesicles. For plasmonic vesicles assembled from AuNRs, they possess tunable absorption in the NIR range, originated from the aspect ratio-dependent LSPR of AuNRs. As an example, Duan *et al.* reported the DOX-loaded GVs assembled from AuNRs grafted with binary PEG/PLA homopolymers.[76] The GVs were designed to be responsive to both NIR laser and enzyme for tumor-targeted controlled delivery of payload.

The large interior space of hollow hybrid vesicles allows for efficient loading of therapeutic agents that can be released in response to stimuli (e.g., pH, enzyme, light or magnetic field), thus achieving effective treatment of tumors. One typical example is the NIR light-triggered release of drugs from nanosized and microsized GVs for imaging-guided cancer therapy.[83,84] The location of nanocarriers, release profile of drugs, and outcomes of treatment can be potentially monitored *in vivo*. Owing to the intrinsic physical properties of inorganic NPs, imaging-guided combination therapy can be developed for the hybrid vesicles, thus maximizing therapeutic efficacy. As an example, GVs loaded with photosensitizer, Ce6 can be used for efficient multimodality imaging-guided combination therapy (i.e., photothermal and photodynamic therapy) of cancers *in vitro* and *in vivo* (Fig. 4.19).[85] Upon 671 nm laser irradiation, the membranes of GVs were disrupted to release the payload of Ce6, leading to the generation of fluorescence emission for imaging and singlet oxygen

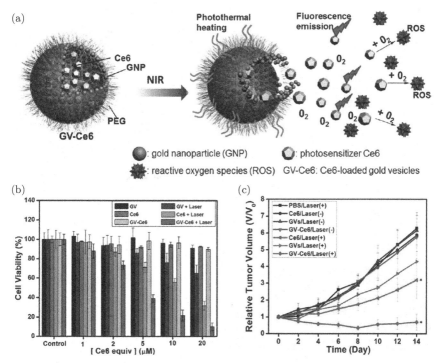

Fig. 4.19. (a) Schematic illustration of utilizing Ce6-loaded GVs for imaging-guided photothermal and photodynamic cancer therapy. (b) *In vitro* study of MDA-MB-435 cell viability for different groups treated with GVs, Ce6, and Ce6-loaded GVs at different concentrations with or without NIR laser irradiation. (c) Tumor growth curves for different groups of tumor-bearing mice treated with GVs, Ce6, and Ce6-loaded GVs with or without NIR irradiation. (Reprinted with permission from Ref. 85, Copyright 2013 American Chemical Society.)

species for killing cancer cells (Fig. 4.19(a)). Simultaneously, the rise in local temperature allowed for photothermal ablation of tumors, as well as photothermal and photoacoustic imaging. In *in vitro* study using MDA-MB-435 cells, this platform showed a 45 to 70% higher therapeutic efficiency than the sum effect of pristine GVs and Ce6 (Fig. 4.19(b)). The same synergistic effect of combined therapies was also observed in *in vivo* studies using breast tumor-bearing mice model. The tumor progression was remarkable delayed for groups administrated with Ce6-loaded GVs and irradiated with laser, compared with control groups including free Ce-6 and unloaded GVs with or without laser irradiation on day 14 (Fig. 4.19(c)).

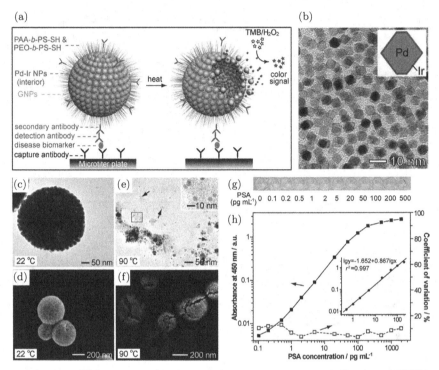

Fig. 4.20. (a) Schematic illustration of utilizing Pd−Ir NPs@GVs based ELISA for detection of disease biomarkers. The Pd−Ir NPs released from captured GVs act as effective peroxidase mimics to catalyze chromogenic substrates. (b) TEM image of Pd−Ir NPs. (c–f) Heat-triggered release of Pd−Ir NPs from GVs. Representative TEM and SEM images of the Pd−Ir NPs@GVs treated at different temperatures (marked in each image) for 1 h. Insert in (e) show magnified TEM images of corresponding regions marked by red boxes. Some of the released Pd−Ir NPs are indicated by black arrows. (g) Representative photographs taken from the ELISA of PSA standards. (h) Corresponding calibration curve (■) and imprecision profile (□) of the detection results shown in (g). Note that absorbance of the blank (i.e., 0 pg/mL PSA) was subtracted from those of PSA standards. Inset shows the linear range region of the calibration curve. (Reprinted with permission from Ref. 87.)

In addition to applications in bioimaging and therapy, hybrid vesicles can be used to develop new generation of tools for cancer diagnosis by taking the advantages of their tailored surface chemistry and hydrophilic interior 3D cavity. As an example, GVs have been integrated in enzyme-based colorimetric assays for signal amplification in detecting disease biomarkers. Recently, Nie and Xia *et al.* developed an enzyme-free signal amplification technique, based on GVs encapsulated with Pd−Ir NPs (Pd−Ir NPs@GVs) as peroxidase mimics,[86] for colorimetric assay of disease biomarkers with significantly enhanced sensitivity. Enzyme-linked immunosorbent assay (ELISA) was used as a model platform for demonstration. The dissociation of GVs captured by analytes liberate the release of thousands of individual Pd−Ir NPs at elevated temperature. The released Pd−Ir NPs acted as peroxidase mimics and generated intense color signal by catalyzing the oxidation of 3,3′,5,5′-tetramethylbenzidine (TMB, a classic HRP substrate) by H_2O_2.[86] Using human prostate surface antigen as a model biomarker, the enzyme-free ELISA showed a detection limit at femtogram/mL level, which is over 10^3-fold lower than that of conventional enzyme-based assay when the same antibodies and similar procedure were used. The enzyme-free signal amplificaiton platform does not rely on enzymes which are prone to temperature fluctuation and storage, which makes them promising for applications in such as early detection of cancers, particularly in resource-limited settings.

4.6. Conclusions and Perspective

This chapter provides a review on the recent progress in the self-assembly of isotropically functionalized HINPs in both the condensed state and aqueous solution, as well as the potential biomedical applications of assembled architectures. The key to the anisotropic self-assembly of these isotropic HINPs is the generation of directional interactions between HINPs by designing the surrounding media (e.g., polymer matrix) or engineering the surface chemistry of HINPs. First, polymer matrix (e.g., thin films of BCPs or supramolecules) or liquid/air (or liquid) interface can be used to assist the assembly of HINPs to form 1D, 2D, or 3D hybrid architectures. The interactions between brushes of neighboring HINPs or between HINPs and phase-separated polymer matrix are usually modulated by varying the σ of polymers or the length mismatching of binary homopolymers on the surface of NPs. Second, the rational design of deformable polymer layers on HINPs composed of BCP or mixed homopolymer tethers is crucial to the self-assembly of HINPs (in analogy to molecular self-assembly) into complex functional structures in selective solvents. It is evidenced that the directional interactions between BCP-grafted NPs arise from the redistribution and conformation change of long, flexible polymer tethers, while the lateral phase-separation of brushes on NP surfaces is responsible for the assembly of HINPs with binary immiscible homopolymers. For HINPs decorated with amphiphilic BCP brushes, their self-assembly produced a variety of hybrid structures, such as vesicles with a monolayer of hexagonally packed NPs in the membranes and with controlled sizes, shape (e.g. spherical, hemispherical, disk-like) and morphology (e.g. patchy, Janus-like, chain in membrane).

Despite promising potential, challenges still remain at this frontier. To achieve a predictive framework for the self-assembly of HINPs into target architectures, the underlying mechanisms of HINP assembly at multiple length scales need to be understood better. This requires elegantly designed experiments, complemented by theoretical modeling and simulations. To meet the increasing demand for the construction of more complex architectures, concurrent self-assembly of multiple types of HINPs might be a good choice.[78] However, this will also pose new challenges due to the complexity and multiple length scale features of forces involved in the assembly. In order to tackle this issue, new advances in computational modelling (e.g., newer models and algorithms) are needed.

Current studies on anisotropic self-assembly of HINPs are still mostly focused on manipulating fairly simple NP interactions (e.g., hydrophobic–hydrophobic). With appropriate polymer design, more complex interactions such as electrostatic interactions, dipole-dipole interactions, host-guest interactions, hydrogen-bonding interactions, metal chelating, etc.[88] can be conceivably integrated into HINP systems to achieve higher complexity or new functionality of assembled materials. For example, Guan *et al.* demonstrated that hydrogen-bonding interactions between polyacrylate amide hairs on SiO_2 NPs significantly enhanced the mechanical and self-healing properties of the assembled superlattice nanocomposites.[89] Moreover, depending on specific shape and compositional features of inorganic NP cores, external fields (e.g. shear field, electric and magnetic fields) or templates can be applied to achieve more exquisite control over anisotropic assembly of HINPs.[49,90] Finally, the combination of engineered interparticle interactions and externally applied forces could potentially enable precise hierarchical self-assembly of HINPs into a wealth of functional nanocomposites.[50]

References

1. Nie, Z., Petukhova, A. and Kumacheva, E. (2010). Properties and emerging applications of self-assembled structures made from inorganic nanoparticles, *Nat. Nano* **5**, 15–25.
2. Boles, M. A., Engel, M. and Talapin, D. V. (2016). Self-assembly of colloidal nanocrystals: From intricate structures to functional materials, *Chemical Reviews* **116**, 11220–11289.
3. Tan, S. J., Campolongo, M. J., Luo, D. and Cheng, W. (2011). Building plasmonic nanostructures with DNA, *Nat. Nanotechnol.* **6**, 268–276.
4. Moffitt, M. G. (2013). Self-assembly of polymer brush-functionalized inorganic nanoparticles: From hairy balls to smart molecular mimics, *J. Phys. Chem. Lett.* **4**, 3654–3666.
5. Kalsin, A. M., Fialkowski, M., Paszewski, M., Smoukov, S. K., Bishop, K. J. and Grzybowski, B. A. (2006). Electrostatic self-assembly of binary nanoparticle crystals with a diamond-like lattice, *Science* **312**, 420–424.
6. Mai, Y. and Eisenberg, A. (2012). Self-assembly of block copolymers, *Chem. Soc. Rev.* **41**, 5969–5985.
7. Fernandes, N. J., Koerner, H., Giannelis, E. P. and Vaia, R. A. (2013). Hairy nanoparticle assemblies as one-component functional polymer nanocomposites: Opportunities and challenges, *Mrs Communications* **3**, 13–29.
8. Li, C. Y., Zhao, B. and Zhu, L. (2014). Hairy particles: Theory, synthesis, behavior, and applications, *J. Polym. Sci. Pol. Phys.* **52**, 1581–1582.
9. Zhang, Horsch, M. A., Lamm, M. H. and Glotzer, S. C. (2003). Tethered nano building blocks: Toward a conceptual framework for nanoparticle self-assembly, *Nano Lett.* **3**, 1341–1346.
10. Glotzer, S. C. and Solomon, M. J. (2007). Anisotropy of building blocks and their assembly into complex structures, *Nature Materials* **6**, 557–562.
11. Zhang, W.-B., Yu, X., Wang, C.-L., Sun, H.-J., Hsieh, I. F., Li, Y., Dong, X.-H., Yue, K., Van Horn, R. and Cheng, S. Z. D. (2014). Molecular nanoparticles are unique elements for macromolecular science: From "Nanoatoms" to giant molecules, *Macromolecules* **47**, 1221–1239.
12. Lin, Z., Yang, X., Xu, H., Sakurai, T., Matsuda, W., Seki, S., Zhou, Y., Sun, J., Wu, K. Y., Yan, X. Y., Zhang, R., Huang, M., Mao, J., Wesdemiotis, C., Aida, T., Zhang, W. and Cheng, S. Z. D. (2017). Topologically directed assemblies of semiconducting sphere-rod conjugates, *J. Am. Chem. Soc.* **139**, 18616–18622.
13. Nie, Z., Fava, D., Kumacheva, E., Zou, S., Walker, G. C. and Rubinstein, M. (2007). Self-assembly of metal-polymer analogues of amphiphilic triblock copolymers, *Nat. Mater.* **6**, 609–614.

14. Grzelczak, M., Sanchez-Iglesias, A., Mezerji, H. H., Bals, S., Perez-Juste, J. and Liz-Marzan, L. M. (2012). Steric hindrance induces crosslike self-assembly of gold nanodumbbells, *Nano Lett.* **12**, 4380–4384.

15. Song, J., Wu, B., Zhou, Z., Zhu, G., Liu, Y., Yang, Z., Lin, L., Yu, G., Zhang, F., Zhang, G., Duan, H., Stucky, G. D. and Chen, X. (2017). Double-layered plasmonic-magnetic vesicles by self-assembly of Janus amphiphilic gold-iron(II,III) oxide nanoparticles, *Angew. Chem. Int. Ed. Engl.* **56**, 8110–8114.

16. Liu, K., Nie, Z., Zhao, N., Li, W., Rubinstein, M. and Kumacheva, E. (2010). Step-growth polymerization of inorganic nanoparticles, *Science* **329**, 197–200.

17. Wang, B., Li, B., Zhao, B. and Li, C. Y. (2008). Amphiphilic Janus gold nanoparticles via combining "Solid-state grafting-to" and "grafting-from" methods, *J. Am. Chem. Soc.* **130**, 11594–11595.

18. Wang, Y., Wang, Y., Breed, D. R., Manoharan, V. N., Feng, L., Hollingsworth, A. D., Weck, M. and Pine, D. J. (2012). Colloids with valence and specific directional bonding, *Nature* **491**, 51–55.

19. Akcora, P., Liu, H., Kumar, S. K., Moll, J., Li, Y., Benicewicz, B. C., Schadler, L. S., Acehan, D., Panagiotopoulos, A. Z., Pryamitsyn, V., Ganesan, V., Ilavsky, J., Thiyagarajan, P., Colby, R. H. and Douglas, J. F. (2009). Anisotropic self-assembly of spherical polymer-grafted nanoparticles, *Nat. Mater.* **8**, 354–359.

20. Choueiri, R. M., Galati, E., Therien-Aubin, H., Klinkova, A., Larin, E. M., Querejeta-Fernandez, A., Han, L., Xin, H. L., Gang, O., Zhulina, E. B., Rubinstein, M. and Kumacheva, E. (2016). Surface patterning of nanoparticles with polymer patches, *Nature* **538**, 79–83.

21. Chen, L. and Klok, H. A. (2013). "Multifaceted" polymer coated, gold nanoparticles, *Soft. Matter.* **9**, 10678–10688.

22. Song, J., Huang, P., Duan, H. and Chen, X. (2015). Plasmonic vesicles of amphiphilic nanocrystals: Optically active multifunctional platform for cancer diagnosis and therapy, *Acc. Chem. Res.* **48**, 2506–2515.

23. Lenart, W. R. and Hore, M. J. A. (2017). Structure–property relationships of polymer-grafted nanospheres for designing advanced nanocomposites, *Nano-Structures & Nano-Objects*

24. Hore, M. J. A. and Composto, R. J. (2014). Functional polymer nanocomposites enhanced by nanorods, *Macromolecules* **47**, 875–887.

25. Yi, C., Zhang, S., Webb, K. T. and Nie, Z. (2017). Anisotropic self-assembly of hairy inorganic nanoparticles, *Acc. Chem. Res.* **50**, 12–21.

26. Dukes, D., Li, Y., Lewis, S., Benicewicz, B., Schadler, L. and Kumar, S. K. (2010). Conformational transitions of spherical polymer brushes: Synthesis, characterization, and theory, *Macromolecules* **43**, 1564–1570.

27. Chiu, J. J., Kim, B. J., Kramer, E. J. and Pine, D. J. (2005). Control of nanoparticle location in block copolymers, *J. Am. Chem. Soc.* **127**, 5036–5037.

28. Guo, Y., Harirchian-Saei, S., Izumi, C. M. S. and Moffitt, M. G. (2011). Block copolymer mimetic self-assembly of inorganic nanoparticles, *ACS Nano* **5**, 3309–3318.

29. Pang, X., He, Y., Jung, J. and Lin, Z. (2016). 1D nanocrystals with precisely controlled dimensions, compositions, and architectures, *Science* **353**, 1268–1272.

30. Pang, X., Zhao, L., Han, W., Xin, X. and Lin, Z. (2013). A general and robust strategy for the synthesis of nearly monodisperse colloidal nanocrystals, *Nat. Nano* **8**, 426–431.

31. Hui, C. M., Pietrasik, J., Schmitt, M., Mahoney, C., Choi, J., Bockstaller, M. R. and Matyjaszewski, K. (2014). Surface-initiated polymerization as an enabling tool for multifunctional (Nano-)engineered hybrid materials, *Chem. Mater.* **26**, 745–762.

32. Song, J., Cheng, L., Liu, A., Yin, J., Kuang, M. and Duan, H. (2011). Plasmonic vesicles of amphiphilic gold nanocrystals: Self-assembly and external-stimuli-triggered destruction, *J. Am. Chem. Soc.* **133**, 10760–10763.

33. Song, J., Zhou, J. and Duan, H. (2012). Self-assembled plasmonic vesicles of SERS-encoded amphiphilic gold nanoparticles for cancer cell targeting and traceable intracellular drug delivery, *J. Am. Chem. Soc.* **134**, 13458–13469.

34. Zhao, B. and Zhu, L. (2009). Mixed polymer brush-grafted particles: A new class of environmentally responsive nanostructured materials, *Macromolecules* **42**, 9369–9383.

35. Tang, S., Lo, T.-Y., Horton, J. M., Bao, C., Tang, P., Qiu, F., Ho, R.-M., Zhao, B. and Zhu, L. (2013). Direct visualization of three-dimensional morphology in hierarchically self-assembled mixed poly(tert-butyl acrylate)/polystyrene brush-grafted silica nanoparticles, *Macromolecules* **46**, 6575–6584.

36. Bao, C., Tang, S., Wright, R. A. E., Tang, P., Qiu, F., Zhu, L. and Zhao, B. (2014). Effect of molecular weight on lateral microphase separation of mixed homopolymer brushes grafted on silica particles, *Macromolecules* **47**, 6824–6835.

37. Li, W., Bao, C., Wright, R. A. E. and Zhao, B. (2014). Synthesis of mixed poly(ε-caprolactone)/polystyrene brushes from Y-initiator-functionalized silica particles by surface-initiated ring-opening polymerization and nitroxide-mediated radical polymerization, *RSC Adv.* **4**, 18772–18781.

38. Fox, T. L., Tang, S., Horton, J. M., Holdaway, H. A., Zhao, B., Zhu, L. and Stewart, P. L. (2015). In situ characterization of binary mixed polymer brush-grafted silica nanoparticles in aqueous and organic solvents by cryo-electron tomography, *Langmuir.* **31**, 8680–8688.

39. Tang, S., Fox, T. L., Lo, T.-Y., Horton, J. M., Ho, R.-M., Zhao, B., Stewart, P. L. and Zhu, L. (2015). Environmentally responsive self-assembly of mixed poly(tert-butyl acrylate)-polystyrene brush-grafted silica nanoparticles in selective polymer matrices, *Soft. Matter.* **11**, 5501–5512.

40. Zubarev, E. R., Xu, J., Sayyad, A. and Gibson, J. D. (2006). Amphiphilicity-driven organization of nanoparticles into discrete assemblies, *J. Am. Chem. Soc.* **128**, 15098–15099.

41. Hu, J., Wu, T., Zhang, G. and Liu, S. (2012). Efficient synthesis of single gold nanoparticle hybrid amphiphilic triblock copolymers and their controlled self-assembly, *J. Am. Chem. Soc.* **134**, 7624–7627.

42. He, J., Liu, Y., Babu, T., Wei, Z. and Nie, Z. (2012). Self-assembly of inorganic nanoparticle vesicles and tubules driven by tethered linear block copolymers, *J. Am. Chem. Soc.* **134**, 11342–11345.

43. Huang, P., Lin, J., Li, W., Rong, P., Wang, Z., Wang, S., Wang, X., Sun, X., Aronova, M., Niu, G., Leapman, R. D., Nie, Z. and Chen, X. (2013). Biodegradable gold nanovesicles with an ultrastrong plasmonic coupling effect for photoacoustic imaging and photothermal therapy, *Angew. Chem. Int. Ed.* **52**, 13958–13964.

44. Palui, G., Aldeek, F., Wang, W. and Mattoussi, H. (2015). Strategies for interfacing inorganic nanocrystals with biological systems based on polymer-coating, *Chem. Soc. Rev.* **44**, 193–227.

45. Ye, X., Zhu, C., Ercius, P., Raja, S. N., He, B., Jones, M. R., Hauwiller, M. R., Liu, Y., Xu, T. and Alivisatos, A. P. (2015). Structural diversity in binary superlattices self-assembled from polymer-grafted nanocrystals, *Nat. Commun.* **6**, 10052.

46. Liu, Y., He, J., Yang, K., Yi, C., Liu, Y., Nie, L., Khashab, N. M., Chen, X. and Nie, Z. (2015). Folding up of gold nanoparticle strings into plasmonic vesicles for enhanced photoacoustic imaging, *Angew. Chem. Int. Ed.* **54**, 15809–15812.

47. Bishop, K. J. (2016). Hierarchical self-assembly for nanomedicine, *Angew. Chem. Int. Ed.* **55**, 1598–1600.

48. Asai, M., Zhao, D. and Kumar, S. K. (2017). Role of grafting mechanism on the polymer coverage and self-assembly of hairy nanoparticles, *ACS Nano* **11**, 7028–7035.

49. Wang, K., Jin, S. M., Xu, J., Liang, R., Shezad, K., Xue, Z., Xie, X., Lee, E. and Zhu, J. (2016). Electric-field-assisted assembly of polymer-tethered gold nanorods in cylindrical nanopores, *ACS Nano* **10**, 4954–4960.

50. Min, Y., Akbulut, M., Kristiansen, K., Golan, Y. and Israelachvili, J. (2008). The role of interparticle and external forces in nanoparticle assembly, *Nat. Mater.* **7**, 527–538.

51. Si, K. J., Sikdar, D., Chen, Y., Eftekhari, F., Xu, Z., Tang, Y., Xiong, W., Guo, P., Zhang, S., Lu, Y., Bao, Q., Zhu, W., Premaratne, M. and Cheng, W. (2014). Giant plasmene nanosheets, nanoribbons, and origami, *ACS Nano* **8**, 11086–11093.

52. Ohno, K., Morinaga, T., Takeno, S., Tsujii, Y. and Fukuda, T. (2007). Suspensions of silica particles grafted with concentrated polymer brush: Effects of graft chain length on brush layer thickness and colloidal crystallization, *Macromolecules* **40**, 9143–9150.

53. Balazs, A. C., Emrick, T. and Russell, T. P. (2006). Nanoparticle polymer composites: Where two small worlds meet, *Science* **314**, 1107–1110.

54. Lin, Y., Boker, A., He, J., Sill, K., Xiang, H., Abetz, C., Li, X., Wang, J., Emrick, T., Long, S., Wang, Q., Balazs, A. and Russell, T. P. (2005). Self-directed self-assembly of nanoparticle/copolymer mixtures, *Nature* **434**, 55–59.

55. Gao, B., Arya, G. and Tao, A. R. (2012). Self-orienting nanocubes for the assembly of plasmonic nanojunctions, *Nat. Nanotechnol.* **7**, 433–437.

56. Nepal, D., Onses, M. S., Park, K., Jespersen, M., Thode, C. J., Nealey, P. F. and Vaia, R. A. (2012). Control over position, orientation, and spacing of arrays of gold nanorods using chemically nanopatterned surfaces and tailored particle-particle-surface interactions, *ACS Nano* **6**, 5693–5701.

57. Nakano, T., Kawaguchi, D. and Matsushita, Y. (2013). Anisotropic self-assembly of gold nanoparticle grafted with polyisoprene and polystyrene having symmetric polymer composition, *J. Am. Chem. Soc.* **135**, 6798–6801.

58. Gao, B., Alvi, Y., Rosen, D., Lav, M. and Tao, A. R. (2013). Designer nanojunctions: Orienting shaped nanoparticles within polymer thin-film nanocomposites, *Chem. Comm.* **49**, 4382–4384.

59. Jiang, G., Hore, M. J., Gam, S. and Composto, R. J. (2012). Gold nanorods dispersed in homopolymer films: Optical properties controlled by self-assembly and percolation of nanorods, *ACS Nano* **6**, 1578–1588.

60. Liu, Y., Rafailovich, M. H., Sokolov, J., Schwarz, S. A., Zhong, X., Eisenberg, A., Kramer, E. J., Sauer, B. B. and Satija, S. (1994). Wetting behavior of homopolymer films on chemically similar block copolymer surfaces, *Phys. Rev. Lett.* **73**, 440–443.

61. Cheyne, R. B. and Moffitt, M. G. (2005). Hierarchical nanoparticle/block copolymer surface features via synergistic self-assembly at the air-water interface, *Langmuir.* **21**, 10297–10300.

62. Li, W., Liu, S., Deng, R. and Zhu, J. (2011). Encapsulation of nanoparticles in block copolymer micellar aggregates by directed supramolecular assembly, *Angew. Chem. Int. Ed.* **50**, 5865–5868.

63. Kao, J. and Xu, T. (2015). Nanoparticle assemblies in supramolecular nanocomposite thin films: Concentration dependence, *J. Am. Chem. Soc.* **137**, 6356–6365.

64. Li, W. K., Liu, S. Q., Deng, R. H., Wang, J. Y., Nie, Z. H. and Zhu, J. T. (2013). A simple route to improve inorganic nanoparticles loading efficiency in block copolymer micelles, *Macromolecules* **46**, 2282–2291.

65. Li, W. K., Zhang, P., Dai, M., He, J., Babu, T., Xu, Y. L., Deng, R. H., Liang, R. J., Lu, M. H., Nie, Z. H. and Zhu, J. T. (2013). Ordering of gold nanorods in confined spaces by directed assembly, *Macromolecules* **46**, 2241–2248.

66. Edgecombe, S. R., Gardiner, J. M. and Matsen, M. W. (2002). Suppressing autophobic dewetting by using a bimodal brush, *Macromolecules* **35**, 6475–6477.

67. Li, W., Wang, K., Zhang, P., He, J., Xu, S., Liao, Y., Zhu, J., Xie, X. and Nie, Z. (2016). Self-assembly of shaped nanoparticles into free-standing 2D and 3D superlattices, *Small* **12**, 499–505.

68. Yan, N., Zhang, Y., He, Y., Zhu, Y. and Jiang, W. (2017). Controllable location of inorganic nanoparticles on block copolymer self-assembled scaffolds by tailoring the entropy and enthalpy contributions, *Macromolecules* **50**, 6771–6778.

69. Deng, R., Li, H., Liang, F., Zhu, J., Li, B., Xie, X. and Yang, Z. (2015). Soft colloidal molecules with tunable geometry by 3D confined assembly of block copolymers, *Macromolecules* **48**, 5855–5860.

70. Deng, R., Li, H., Zhu, J., Li, B., Liang, F., Jia, F., Qu, X. and Yang, Z. (2016). Janus nanoparticles of block copolymers by emulsion solvent evaporation induced assembly, *Macromolecules* **49**, 1362–1368.

71. Hsu, C. H., Dong, X. H., Lin, Z., Ni, B., Lu, P., Jiang, Z., Tian, D., Shi, A. C., Thomas, E. L. and Cheng, S. Z. (2016). Tunable affinity and molecular architecture lead to diverse self-assembled supramolecular structures in thin films, *ACS Nano* **10**, 919–929.

72. Koerner, H., Drummy, L. F., Benicewicz, B., Li, Y. and Vaia, R. A. (2013). Nonisotropic self-organization of single-component hairy nanoparticle assemblies, *ACS Macro. Lett.* **2**, 670–676.

73. Percebom, A. M., Giner-Casares, J. J., Claes, N., Bals, S., Loh, W. and Liz-Marzan, L. M. (2016). Janus gold nanoparticles obtained via spontaneous binary polymer shell segregation, *Chem. Commun.* **52**, 4278–4281.

74. Wang, Y., Yang, G., Tang, P., Qiu, F., Yang, Y. and Zhu, L. (2011). Mixed homopolymer brushes grafted onto a nanosphere, *J. Chem. Phys.* **134**, 134903.

75. Lee, J. C. M., Bermudez, H., Discher, B. M., Sheehan, M. A., Won, Y.-Y., Bates, F. S. and Discher, D. E. (2001). Preparation, stability, and in vitro performance of vesicles made with diblock copolymers, *Biotech. Bioengin.* **73**, 135–145.

76. Song, J., Pu, L., Zhou, J., Duan, B. and Duan, H. (2013). Biodegradable theranostic plasmonic vesicles of amphiphilic gold nanorods, *ACS Nano* **7**, 9947–9960.

77. He, J., Huang, X., Li, Y. C., Liu, Y., Babu, T., Aronova, M. A., Wang, S., Lu, Z., Chen, X. and Nie, Z. (2013). Self-assembly of amphiphilic plasmonic micelle-like nanoparticles in selective solvents, *J. Am. Chem. Soc.* **135**, 7974–7984.

78. Liu, Y. J., Liu, B. and Nie, Z. H. (2015). Concurrent self-assembly of amphiphiles into nanoarchitectures with increasing complexity, *Nano Today* **10**, 278–300.

79. Liu, Y., Li, Y., He, J., Duelge, K. J., Lu, Z. and Nie, Z. (2014). Entropy-driven pattern formation of hybrid vesicular assemblies made from molecular and nanoparticle amphiphiles, *J. Am. Chem. Soc.* **136**, 2602–2610.

80. Liu, Y., Yang, X., Huang, Z., Huang, P., Zhang, Y., Deng, L., Wang, Z., Zhou, Z., Liu, Y., Kalish, H., Khachab, N. M., Chen, X. and Nie, Z. (2016). Magneto-plasmonic Janus vesicles for magnetic field-enhanced photoacoustic and magnetic resonance imaging of tumors, *Angew. Chem. Int. Ed. Engl.* **55**, 15297–15300.

81. Wang, L., Liu, Y., He, J., Hourwitz, M. J., Yang, Y., Fourkas, J. T., Han, X. and Nie, Z. (2015). Continuous microfluidic self-assembly of hybrid Janus-like vesicular motors: Autonomous propulsion and controlled release, *Small* **11**, 3762–3767.

82. He, J., Wang, L., Wei, Z., Yang, Y., Wang, C., Han, X. and Nie, Z. (2013). Vesicular self-assembly of colloidal amphiphiles in microfluidics, *ACS Appl. Mater. Interfaces* **5**, 9746–9751.

83. He, J., Wei, Z., Wang, L., Tomova, Z., Babu, T., Wang, C., Han, X., Fourkas, J. T. and Nie, Z. (2013). Hydrodynamically driven self-assembly of giant vesicles of metal nanoparticles for remote-controlled release, *Angew. Chem. Int. Ed.* **52**, 2463–2468.

84. He, J., Zhang, P., Babu, T., Liu, Y., Gong, J. and Nie, Z. (2013). Near-infrared light-responsive vesicles of Au nanoflowers, *Chem. Commun.* **49**, 576–578.

85. Lin, J., Wang, S., Huang, P., Wang, Z., Chen, S., Niu, G., Li, W., He, J., Cui, D., Lu, G., Chen, X. and Nie, Z. (2013). Photosensitizer-loaded gold vesicles with strong plasmonic coupling effect for imaging-guided photothermal/photodynamic therapy, *ACS Nano* **7**, 5320–5329.

86. Xia, X., Zhang, J., Lu, N., Kim, M. J., Ghale, K., Xu, Y., McKenzie, E., Liu, J. and Ye, H. (2015). Pd–Ir core–shell nanocubes: A type of highly efficient and versatile peroxidase mimic, *ACS Nano* **9**, 9994–10004.

87. Ye, H., Yang, K., Tao, J., Liu, Y., Zhang, Q., Habibi, S., Nie, Z. and Xia, X. (2017). An enzyme-free signal amplification technique for ultrasensitive colorimetric assay of disease biomarkers, *ACS Nano* **11**, 2052–2059.

88. Elacqua, E., Zheng, X., Shillingford, C., Liu, M. and Weck, M. (2017). Molecular recognition in the colloidal world, *Acc. Chem. Res.* **50**, 2756–2766.

89. Williams, G. A., Ishige, R., Cromwell, O. R., Chung, J., Takahara, A. and Guan, Z. (2015). Mechanically robust and self-healable superlattice nanocomposites by self-assembly of single-component "sticky" polymer-grafted nanoparticles, *Adv. Mater.* **27**, 3934–3941.

90. Singh, G., Chan, H., Baskin, A., Gelman, E., Repnin, N., Kral, P. and Klajn, R. (2014). Self-assembly of magnetite nanocubes into helical superstructures, *Science* **345**, 1149–1153.

CHAPTER 5

Surface Coated NIR Light-Responsive Nanostructures for Imaging and Therapeutic Applications

YANMEI YANG[*,‡] and **BENGANG XING**[†,§,¶]

[*]College of Chemistry, Chemical Engineering and Materials Science,
Key Laboratory of Molecular and Nano Probes,
Ministry of Education, Collaborative Innovation Center
of Functionalized Probes for Chemical Imaging in
Universities of Shandong, Institute of Molecular and Nano Science,
Shandong Normal University, Jinan 250014, China
[†]Division of Chemistry and Biological Chemistry,
School of Physical and Mathematical Sciences,
Nanyang Technological University,
Singapore, 637371, Singapore
[§]Sino-Singapore International Joint Research Institute (SSIJRI),
Guangzhou 510000, China

5.1. Introduction

Currently, targeted disease therapy without affecting healthy tissues or cells remains a clinical concern in hospitals and healthcare community. One prerequisite requirement to achieve precise targeting

[‡]Corresponding author: yym@suda.edu.cn
[¶]Corresponding author: bengang@ntu.edu.sg

localization is that sufficient therapeutic agents should be capable of selectively delivering at the specific site of diseases. However, most existing drug molecules and therapeutic agents do not have unique selectivity to the diseases areas and they could not accumulate at the disease areas automatically, and thereby, the rational design with optimal performance by effectively increasing the targeted therapeutic agent concentration, while minimizing systematic side effects will be highly significant in clinical practice and extensive efforts still need to be well developed.[1,2] So far, numerous approaches have been considered to achieve the targeted therapy by selectively directing the drug molecules at the disease areas.[3–6] Thanks for the unique advantages of nanotechnology, such accurate release of therapeutic agents at the disease site for selective killing of malicious cells without side-effect ones can be easily accomplished. Until now, many nanocarriers can be used to transport drug molecules to the disease area.[7] The non-invasive strategy, which utilizes the external stimuli to control the delivery of therapeutic payloads from the nanostructures in the desired location is an ideal therapeutic method. Among the numerous strategies demonstrated, those based on the light or photo have garnered huge interest since light as an external stimulus is not only non-invasive but also enables a control release of therapeutic agents controllable both spatially and temporally.[8]

So far, the development of light-controlled bioimaging and therapies have obtained enormous attention, which are mainly attributed to the reason that light offers a remote activation of a wide range of materials at a specific time and location with relatively high precision.[9–12] Generally, it is widely accepted that the short wavelength light displayed poor tissue penetration capability, potential phototoxicity and high auto fluorescence background, which thus limit its further applications for therapy *in vivo*.[13–15] Therefore, the development of long wavelength light-induced bioimaging and therapeutic agents delivery was demonstrated to be a promising option.[16–18] Near-infrared (NIR) light would be much more desirable than short wavelength light (e.g. UV or visible light) in nanomedicine due to its superior penetration ability for remote photoactivation of materials at a specific time and location with relatively low interference and high precision. As a consequence, for the *in vivo* therapies of tumors under skin and deeply seated within tissue, NIR light can penetrate the tissue to reach the deeply placed optical-sensitive agents.[19]

A variety of promising strategies have been demonstrated for the targeted therapies by using NIR light-responsive nanostructures.[18,20–22] In order to achieve such purpose, one of the commonly used approaches is the case to make use of photosensitive nanoplatform itself, which can convert NIR light to short wavelength emissions or to heat for thermal-responsive delivery of the drug active molecules.[23] Normally, the payload on the nanostructure surface or encapsulated within nanocarriers will be systematically released when irradiation with a special wavelength light.[24,25] The second well-established approach is to utilize the nanostructures which can absorb NIR light and followed by converting the photo energy into localized thermal energy.[26] The heat transduced can be used to kill the abnormal cells nearby the nanomaterials, which is known as the photothermal therapy (PTT). The third commonly applied strategy is combining the nanomaterials with photosensitizers together. When upon NIR light illumination, the nanocomposites can generate reactive oxygen species (ROS), such as singlet oxygen ($^1O_2$), so that can be used for photodynamic therapy (PDT) application.[27] By taking the advantages of NIR light-assisted therapies, the photosensitive nanostructures which are NIR-responsive have raised tremendous interest in the area of biomedical research. Currently, many kinds of nanomaterials have been proposed for the NIR light-controlled therapies, including gold-based nanomaterials, upconversion nanoparticles (UCNPs), carbon nanotubes (CNTs) and some two-dimensional (2D) nanomaterials materials.[19,28] In this chapter, we will mainly focus on reviewing the imaging and therapeutic applications of these nanomaterials with surface coated, through the following subheadings, surface coated gold-based nanomaterials, surface coated UCNPs, surface coated carbon nanotubes and surface coated 2D nanomaterials.

5.2. Surface Coated Gold-based Nanomaterials

Gold-based nanostructures, such as gold nanorods (AuNRs), gold nanoshells (AuNSs), assembled gold nanoparticles, exhibit biocompatibility, easy surface functionalization, tunable physical properties and strong NIR absorbance, which provide a versatile platform for a board range of applications in biomedical applications.[29–31] These noble metal gold-related nanostructures are light responsive mainly due to the localized surface plasma resonance (LSPR), which is the intrinsic property of gold. In this regard, gold related nanomaterials will be very suitable platforms for NIR light-controlled therapy.

5.2.1. *Drug delivery*

As mentioned above, AuNRs own strong absorption in the NIR region due to their longitudinal surface plasma resonance. However, for the bare AuNRs without any stabilizing ligands or surface coating, they will be unstable and easily aggregation in living condition.[32] There have been various approaches to conjugate drug molecules with AuNRs for NIR-controlled delivery. For example, Qu and co-workers have developed a NIR light-triggered drug delivery system based on AuNRs.[33] In order to enhance the loading efficiency of drug molecules and improve the stability of AuNRs conjugates in biological environment, AuNRs were coated with mesoporous silica shell. The as-obtained core-shell NPs were surface-modified with DNA aptamer moiety for effective capping of fluorophore (FITC) and antitumor drug, doxorubicin (Dox). Upon NIR light irradiation, the local temperature would be raised due to the photothermal effect of AuNRs, resulting in the opening of aptamer cover and thus releasing of entrapped drug molecules. Similarly, another related work, which also focused on the design of NIR light-to-thermal responsive drug delivery, was demonstrated by the Yeh's group.[34] Besides DNA aptamer moieties, Chen *et al.* also employed a kind of thermo- and pH-responsive polymers to functionalize the AuNRs.[35] AuNRs were first coated with mesoporous silica shell for loading drug molecules and the obtained $Au@SiO_2$ nanoparticles were further encapsulated into the thermo- and pH-responsive polymer which mainly consisting of poly (N-isopropylacrylamide) (pNIPAm) (Fig. 5.1). The temperature of the nanocomposites will rapidly increase when irradiated with NIR light, thereby the drug release rate can be easily

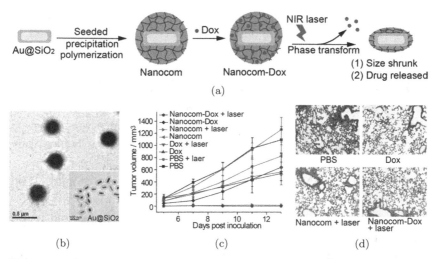

Fig. 5.1. a) Schematic illustration of the synthesis of pNIPAm coated Au@SiO$_2$ nanocomposite; b) TEM images of the pNIPAm coated Au@SiO$_2$ nanocomposite (inset: TEM of Au@SiO$_2$); c) The the tumor volume curve of the four laser-irradiated groups. The antitumor activity including PBS, Dox, Nanocomposite, and Nanocom-Dox groups (with or without laser irradiation) through tail vein; d) HE staining of lung tissues after treatment and red arrows indicate the metastases. Reproduced with permission from Ref. 35. Copyright 2014, American Chemical Society.

improved by the generated heat. In this work, the mechanism of the light controlled drug release strategy is mostly attributed to the laser converted thermal effect dissociated electrostatic interactions between drug molecules of Dox and the polymer shell structures.

AuNSs are usually have the morphology that either a hollow gold shell or a composite particle comprising a core (such as a silica seed) packaged by a gold shell, which can serve as NIR-controlled drug delivery platforms because of their excellent NIR-absorption capability. Several novel designs for NIR-controlled drug delivery on the basis of AuNSs have been demonstrated.[20,36,37] By taking the polyethylene glycol (PEG) coated-hollow gold nanoshells (HAuNSs) as example, the anticancer drug Dox was loaded on its surface through electrostatic adsorption.[38] Upon exposed with 808 nm laser (5 W/cm^2), the absorption intensity of Dox in aqueous solution of Dox@HAuNSs was found to be increased significantly, which indicating the effective release of free drug molecules from nanoplatforms. This NIR light responsive nanocarrier can also be used in human breast carcinoma cells. For the cells which were treated with Dox@HAuNSs only, the anticancer drug molecules did not reach the nuclei, means that the Dox were not released from the HAuNSs. However, for the cancer cells treated with Dox@HAuNSs and followed by NIR laser irradiation, the significant red fluorescence was observed in the cell nuclei, clearly suggesting that the drug molecules were released under NIR laser irradiation.

Apart from AuNRs and AuNSs, some other kinds of gold-based nanomaterials can also be coated with the thermosensitive polymers,[39,40] liposomes[41-43] and even biological samples, like red blood cells.[44] Here we take the smart polymers for example, in 2009, Xia's group fabricated a type of smart polymer pNIPAAm coated Au nanocages for the purpose of NIR light-controlled anticancer drug deliver.[45] The as-prepared Au nanocages could efficiently convert the NIR light into thermal energy, resulting in the collapse of pNIPAAm polymer chains. Therefore, the pores on the nanocages are opened and thereby release the loaded therapeutic molecules into surroundings.

5.2.2. *Photothermal therapy*

Among various nanomaterials, gold-based nanostructures are ideal PTT agents due to their strong optical extinction in the visible and NIR regions. Tipically, the absorbed light energy can be converted into heat, which subsequently release into the disease area and followed by destroy the abnormal cells nearby. Additionally, Gold-based nanostructures have been found to own other excellent properties for biomedicine applications, such as high photostability and good biocompatibility. Currently, many approaches for PTT on the basis of gold-related nanomaterials have achieved encouraging therapeutic success which indicates great promise for future clinic application.

The first work that utilizing AuNPs as photothermal sensitizer in PTT was demonstrated by Lin and co-workers in 2003.[46] Then the

light mediated PTT operation showed highly localized photother-
molysis of the targeted cells which treated with AuNPs. Inspired
by this pioneering study, many other groups have reported the
applications of AuNPs in PTT triggered by long wavelength light
illumination.[47–49] Huang *et al.* proposed a novel route to construct a
biodegradable plasmonic gold nanovesicles for both PTT and photoacoustic (PA) studies in 2013.[50] The gold nanovesicles were prepared
by assembling AuNPs tethered with an amphiphilic block copolymer (PEG-b-PCL, BCP), which showed strong absorption in the
NIR region and enhanced photothermal efficiency upon laser irradiation *in vivo*. More importantly, these assembled gold nanovesicles
were gradually collapse into discrete AuNPs under a temperature-
and time-depended manner, which improved the clearance ability
of AuNPs in living systems after PTT treatment. Besides coating
with polymers, AuNPs can also be assembled with other inorganic
nanomaterials for PTT applications. For example, Su *et al.* has
reported a gold nanoparticles-based NIR hyperthermia agents by
decorating AuNPs with silicon nanowires (AuNPs@SiNWs).[51] The as
synthesized AuNPs@SiNWs possesses strong NIR absorbance from
700–1000 nm and displayed excellent heat production when irradiated under 808 nm laser. These hybrid nanomaterials were found
that can inhibit the tumor grows even with as low concentration
as 150 ppm.

AuNRs are also a kind of promising candidate nanomaterials for
PTT, because they can easily synthesized and can indicate strong
optical extinction in the NIR region.[52] AuNRs can be easily prepared
with an aspect ratio of 4 by using seed-mediated method, which show
a promising longitudinal absorbance at approximately 800 nm.[19,26,53]
By taking advantage of this unique property, nanorod materials can
be used to induce thermal ablation when illuminated with NIR light.
Currently, various strategies have been demonstrated by conjugating
AuNRs with biomolecules for PTT treatment. For example, EI-Sayed
and co-workers demonstrated an *in vitro* study of bioimaging and
PTT by using AuNRs conjugated with anti-epidermal growth factor
receptor (anti-EGFR), which can selectively bind to the malignant-
type cells.[54] Otherwise in another system, Van Maltzahn *et al.* used

PEG stabilized AuNRs for photothermal therapy *in vivo*.[55] They found that PEG-AuNRs complex could be dispersed in a variety of solvents and exhibited long circulation half-life *in vivo* ($t_{1/2} \sim 17$ h). Furthermore, this PEG protected AuNRs exhibit superior photothermal efficiency when injected into the tumors of mice. When upon NIR laser irradiation (810 nm, 2 W/cm^2, 5 min), the temperature of the targeted area was rapidly increased to over 70°C. Benefit from the unique and efficient performance of AuNRs to convert NIR light to heat, this platform has achieved encouraging therapeutic efficacies, which successfully annihilated the tumors within 10 days. In another work, Tian's group designed a novel nanoplatform by assembling DNA coated AuNRs on the surface of a DNA-origami structure (D-AuNRs).[56] (Fig. 5.2) In comparison to AuNRs, the obtained D-AuNRs hybrids can function both as an efficient photoacoustic imaging agent for cancer diagnosis and PTT for cancer therapy.

5.2.3. *Photodynamic therapy*

PDT is a unique type of therapeutic modality that relies on photosensitizer (PS) agents to generate reactive oxygen species under light illumination, such as the $^1O_2$, which are usually associated with oxidative stress and subsequently causing cell damage.[16] Normally, three key components are essential in a PDT system: a PS molecule, oxygen and a light source which can excite the PS molecules. Under light irradiation, the generated free radicals from the PS agents will oxidize the biomacromolecule components in cells, like DNA, certain enzymes or proteins, which will induce the cytotoxicity and even kill cells.[57] So far, PDT treatment has been well recognized and approved by the FDA for cancer treatment in clinics. In general, the PS agents employed in conventional PDT are molecules that contain porphyrins or their analog structures. Consequently, most of the PS agents are hydrophobic, which usually have poor water solubility and lack of effective selectivity/accumulation at the target site of diseases. To overcome these obstacles and meanwhile to improve the therapeutic efficacy, many nanocarriers have been proposed to facilitate the targeted PS agent delivery for PDT *in vivo*. Currently,

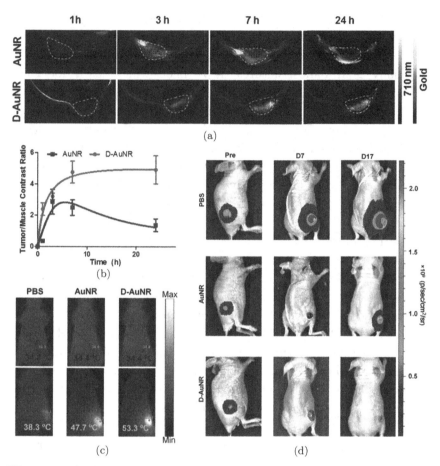

Fig. 5.2. a) Photoacoustic imaging in the tumor site at several different time intervals after intravenous injection of AuNRs and D- AuNRs in 4T1-tumor bearing mice; b) Contrast ratio between the tumor and the region of back muscles extracted from the images for AuNRs and D–AuNRs; c) Infrared thermographic maps of mice subjected to intravenous administration of PBS, AuNRs, and D–AuNRs measured 10 min after NIR irradiation (808 nm, 1.5 W/cm^2); d) Bioluminescence imaging of 4T1-fLuc-tumor bearing mice intravenously injected with PBS in the control group, AuNRs and D–AuNRs, followed by 10 min NIR laser irradiation. Reproduced with permission from Ref. 56. Copyright 2016, John Wiley & Sons, Inc.

extensive efforts have been investigated to explore the feasibility of the PDT bio-applications through long wavelength NIR light irradiation mostly due to its capability for deeper tissue penetration. The usage of NIR light-responsive PDT could significantly improve the safety of PDT treatment, effectively reduce the light-induced cytotoxicity and minimize the possible photosensitivity to ambient light illumination. By right, many kinds of inorganic nanomaterials, especially gold nanostructures, have been widely used for PDT applications.

Many literatures clearly reported that the combination of AuNPs and organic photosensitizers can improve the PDT treatment effects. For example, Cheng *et al.* recently designed a NIR-light triggered PDT by immobilizing photosensitizer, Pc 227, on AuNPs surface via Au-S bond.[58] AuNPs were firstly stabilized with PEG to improve the biocompatibility, followed by loading with Pc 227 molecules. Upon irradiation with NIR light, the AuNP-Pc 227 conjugates showed efficient PDT effect to inhibit the growth of cancer cells. Instead of coating with photosensitizers, some gold nanostructures can directly generate ROS under light irradiation. Vankayala *et al.* presented this interesting concept by using AuNRs alone that can sensitize the formation of singlet oxygen.[59] The cationic lipid coated AuNRs can act as activatable PS agents, which was studied in cellular level. Under NIR light irradiation, this nanoplatform can successfully inhibit the growth of melanoma tumors without any conjugation of external PS agents. This novel lipid coated AuNRs also showed high performance of PDT effect *in vivo*. After administration with lipid coated AuNRs and followed by laser irradiation (915 nm, $130\,\text{mW/cm}^2$), the B16F0 melanoma tumors in mice have been found destroyed completely.

5.3. Surface Coated Upconversion Nanoparticles

As mentioned previously, long wavelength light-mediated drug activation in biomedical applications will be always preferred as they will be less detrimental to tissue, and have deeper light penetration capability. Recently, one of the promising strategies for such NIR light-responsive therapies will be through the application of UCNPs. Lanthanide-doped UCNPs have attracted a tremendous amount of attention in recent years because of their special properties that can convert NIR light irradiation to shorter wavelength emissions (normally, from UV to visible even to NIR region).[60–62] In this section, we will mainly focus on the introduction of UCNPs for their NIR light-controlled drug delivery, PTT and PDT applications.

5.3.1. *Drug delivery*

Because of the special property that can convert NIR light irradiation into shorter wavelength emissions, UCNPs have attracted a tremendous amount of consideration in recent years.[63,64] The upconverted emission can be used to photocleave the "caging" moieties and release the bioactive molecules in a spatial and temporal precision. In 2012, Xing and co-workers demonstrated the concept that NIR light (980 nm)-induced release of caged d-luciferin based on silica coated UCNPs.[65] In this work, solid silica shell were used to coated the UCNPs firstly (UCNPs@SiO$_2$) to facilitate the increased biocompatibility and subsequent particle surface functionalization. For the photocage moieties, the commonly used photoliable "caging" moieties, *o*-nitrobenzyl, were employed to block *d*-luciferin molecules, one robust bioluminescent probe for specific recognition of gene reporter firefly luciferase, were further covalently linked to the NPs surface. Then the two parts were conjugated together through the covalently bond. Once under the NIR laser irradiation, the "caged" *d*-luciferin will be released by one of the converted UV emissions from UCNPs. This uncaging of *d*-luciferin molecules can react with firefly luciferase to produce photon at 560 nm, which can be thus used for bioluminescence imaging *in vitro* and *in vivo*. One similar approach was also carried out to control the release of the anticancer drug 5-fluorouracil, which was demonstrated by Fedoryshin *et al.*[66] On the other side, the upconverted UV emission from NIR laser excited UCNPs can be used for control of drug activities in cancer therapy directly. Based on this design, several antitumor reagents which indicate light-induced cytotoxicity have been used to conjugate with surface modified UCNPs for the targeted tumor inhibition. For example, a novel NIR light-activated drug delivery approach based on silica coated UCNPs (UCNPs@SiO$_2$) was developed Xing's group.[67] In this strategy, the unconverted UV emission from UCNPs@SiO$_2$ could locally activate light responsive Pt(IV) prodrug which generates effective antitumor cytotoxicity *in vitro* and *in vivo*. Moreover, the promising antitumor treatment has also been systematically evaluated by multi-modality imaging analysis. Based on the process of programmed cell death induced by NIR light-mediated prodrug activation, the fluorescence imaging through the cleavable DEVD peptide hydrolyzed by apoptosis activated caspase-3 enzyme has also been applied to evaluate the early stage of tumor therapeutic intervention. Nearly at the same time, Lin and co-workers.[68] designed a similarly strategy by using polymers coated UCNPs. The novel photoactive Pt(IV) prodrug, trans, trans, trans-$[Pt(N_3)_2(NH_3)(py)-(O_2CCH_2CH_2CO_2H)]$ was conjugated with PEI/PEG coated core-shell NaYF$_4$:Yb/Tm@NaGdF$_4$/Yb UCNPs. Such unique prodrug UCNPs nanomedical system exhibited potent *in vivo* antitumor activity under NIR laser irradiation (980 nm). Moreover, due to the positive signal enhancement property of Gd^{3+} ions, this polymer coated UCNPs also be used for tri-modality imaging of tumor status and treatment in living condition. Besides the unique design to spatiotemporally activate metal-based antitumor drugs, currently, similar concept has also been broadly applied to light-mediated release of other conventional antitumor reagents, such as Dox.[69,70]

Mesoporous silica shell coated UCNPs have also been utilized for NIR light controlled drug delivery, due to their promising benefits of biocompatibility and photostability.[8] Zhang *et al.* have demonstrated the utilities of mesoporous silica coated UCNPs (UCNPs@mSiO$_2$) for remotely controlled delivery of "photocaged" siRNA and DNA moieties.[71] In their work, siRNA or plasmid DNA was caged by a photoactive linkage, 4,5-dimethoxy-2-nitroacetophenone (DMNPE). These caged biomolecules were encapsulated into the pores of silica shell with a loading efficiency of 70%. When upon 980 nm laser irradiation, these "caged" nucleic acids can be photo-released by the upconverted UV light, and thus achieve the spatial and temporal regulation of gene silencing both *in vitro* and in living tissues. Benefiting from enhanced tissue penetration and minimum phototoxicity of NIR light, the NIR-to-UV triggered drug delivery nanoplatforms based on UCNPs can serve as a new approach in various fields and arise more research interest.

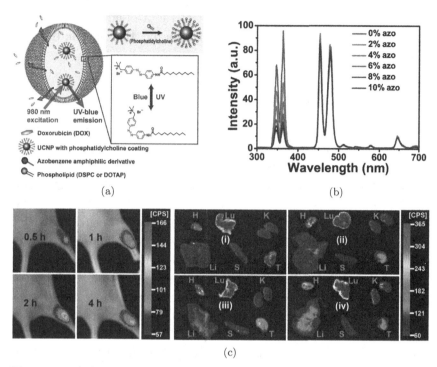

(a)   (b)

(c)

Fig. 5.3. a) Schematic illustration of the azobenzene-liposome/upconversion nanoparticle hybrid vesicles for NIR-triggered drug delivery; b) Emission spectra of UCNP@Azo-Lipo vesicles upon 980 nm NIR light irradiation as a function of different percentage of azobenzene derivative component; c) *In vivo* upconversion luminescence imaging of subcutaneous MCF-7/ADR tumor (right hind leg) borne by nude mice after intravenous injection of Dox loaded UCNP@Azo-Lipo at 0.5, 1, 2, 4 h. Both images were acquired under the same conditions (power density $\approx 200\,\mathrm{mW/cm^2}$); d) *Ex vivo* images of tumors and major organs with Dox fluorescent signals after various treatments: NIR irradiation at different times (0.5 h, 1 h, 2 h, 4 h) post injection, then dissection after 2 h. Reproduced with permission from Ref. 18. Copyright 2016, John Wiley & Sons, Inc.

Besides coated with silica shell, Zhang's group reported a novel NIR light-trigged drug delivery systems by using liposome coated UCNPs.[18] The anticancer drug Dox were encapsulated in azo-benzene doped liposome coated UCNPs hybrid vesicles (UCNPs@Azo-lipo) (Fig. 5.3). Upon NIR light illumination (980 nm), the azo-benzene amphiphilic molecules will transform the *trans*-isomer into *cis*-isomer under the converted UV/vis light from

UCNPs. This reversible isomerization of the azobenzene group will cause the movement of the liposome membrane, therefore induce the drug molecules release.

5.3.2. *Photothermal therapy*

UCNPs are lanthanide-doped nanomaterials which normally cannot generate heat from light due to the low extinction coefficient of the rare-earth components.[72] However, once combination of UCNPs and other plasmonic nanostructures together, such as Au,[73] Ag[74] and Fe_3O_4[75] *et al.*, these hybrid nanocomposites could exhibit high efficiency of thermal generation, which have already been applied for PTT cancer therapy. Besides the plasmonic nanomaterials, other optical absorbing agents including CuS and NIR-absorbing dye molecules have also been reported to integrate with UCNPs for photo-controlled biomedical applications. Shi and co-workers developed a new type of silica-coated UCNPs nanocomposites by surface doped with ultra-small CuS NPs for PTT.[76] In this system, the CuS NPs were employed as photothermal agents to convert light to heat effect. When the solution containing the nanocomposites (1.2 mg/mL) exposed under NIR laser (980 nm, 1.5 W/cm^2, 5 min), the temperature was increased from 25°C to 40°C. The designed PTT agents were intratumorally injecting into the 4T1 cells-bearing mice for 1 h and followed by NIR laser irradiation, the tumor growth was found to be significantly inhibited. This CuS NPs doped silica-coated UCNPs showed the great potential for the promising PTT treatment *in vivo*.

Currently, most of the reported NIR-responsive therapy strategies by using UCNPs were rely on the 980nm laser excitation. However, such long wavelength light may overlap with the water absorption *in vivo*,[77] thus may cause several potential issues, such as overheating to potentially damage biological tissues. Fortunately, by coating UCNPs with NIR-absorbing molecules or doping with new elements, the new types of UCNPs can be excited by the shorter wavelength light, e.g. at 808 nm etc.[78,79] Recently, Liu's group[80] demonstrated a multifunctional BSA coating UCNPs nanoplatform by loading two types of fluorescent dye molecules, the NIR-absorbing dye (IR825) and the photosentitizer dye (Rose Bengal, RB), on the UCNPs' surface. When the nanoplatform irradiated with 808 nm laser for 5 min, the temperature was found to rapidly increase (20°C to 43°C, with the light dosage of 0.5 W/cm^2) *in vivo*, while with the temperature changed from 30°C to 45°C *in vivo* (under the dosage of 0.35 W/cm^2). As shown from their results, this designed multifunctional UCNP-based nanoplatform can be used for imaging-guided cancer therapy.

5.3.3. *Photodynamic therapy*

As mentioned above, UCNPs can convert NIR light into UV or visible light, which could be used to active several kinds of PS molecules, exhibit great promising in NIR light-controlled PDT applications. Zhang and co-workers first demonstrated the *in vitro* PDT application of UCNPs for effectively kill cancer cells.[81] In order to increase their biocompatibility and loading the photosensitizer merocyanine-540 (M540), UCNPs were coated with thin layer of mesoporous silica shell. When the nanocomposites were excited with NIR light (e.g. 974 nm), the PS molecule M540 will produce $^1O_2$ because of the converted UV emission. Furthermore, by conjugating the M540-coated UCNPs antibody, the nanocomposites exhibited the effective cancer targeting and high PDT efficiency for the destruction of cancer cells ability. Since then, many other different groups have studied the UCNP-based PDT applications *in vitro* and *in vivo*.[80,82] For instance Zhang *et al.* designed an *in vivo* PDT approach by using mesoporous silica coated UCNPs (UCNPs@mSiO$_2$) with surface loading by two photosensitizers, which can be activated at the same time by the multicolor-emissions from UCNPs under 980 nm light.[83] This two PS molecules loaded nanoplatform showed improvement of the generation of $^1O_2$ and notably enhance the PDT efficiency. Gu *et al.* also reported an *in vivo* tumor-targeted PDT systems by using zinc(II) phthalocyanine (ZnPc) loaded UCNPs.[84] Typically, UCNPs were coated with folate-modified amphiphilic chitosan (FASOC) and followed by loading hydrophobic ZnPc molecules. The FASOC-UCNP-ZnPc PDT agents were intravenously injected into the s180 tumors-bearing mice and irradiated with 980 nm light after 24 h post-injection ($0.2\,\mathrm{W/cm^2}$, $30\,\mathrm{min}$). The tumor growth was greatly inhibited as compared to the control group in which the mice were only treated with saline.

So far, tumor hypoxia has been recognized as one of the crucial reasons for the poor prognosis of much anticancer therapeutics. To solve this issue, some interesting strategies have been proposed recently. For example, Shi *et al.* demonstrated a double silica-shelled-coated Gd^{3+}-doped UCNPs structure, which were capable of co-delivering PS reagents (e.g., silicon phthalocyanine dihydroxide, SPCD) and a bioreductive prodrug tirapazamine (TPZ).[85] Upon $980\,\mathrm{nm}$ laser exposure, a synergetic *in vivo* cancer therapy could be first achieved under normal oxygen environment, then immediately followed by the induced cytotoxicty of activated TPZ when oxygen is depleted during the PDT process done by UCNPs (Fig. 5.4). Moreover, together with the NIR light mediated therapy, such unique UCNPs platform also provided the possibility for multimodal computed tomography (CT) and positron emission tomography (PET) imaging for the tumor status. The enhanced PDT tumor treatment within deeper tissue, where the environment is highly hypoxic, can be easily achieved.[85–87]

Recently, a new type of Nd^{3+} element doped UCNPs, which can be excited under a NIR wavelength light around $800\,\mathrm{nm}$ light, has been developed.[78,88,89] These shorter NIR wavelength laser near $800\,\mathrm{nm}$ could not only exhibit more robust NIR light-controlled diseases treatment and imaging, meantime, they could also greatly minimize the laser-induced local overheating effect as compared to those done by $980\,\mathrm{nm}$ laser excitation.[79,90,91] Based on this promising advantages, Xing's group demonstrated a novel tumor micro-environment responsive nanoplatform based on UCNPs, with the surface conjugated with protease enzyme sensitive peptide (Fig. 5.5).[92] Upon tumor specific enzyme reactions, the activated peptides sequence could trigger the covalent cross-linking of neighboring particles which specifically induced the self-accumulation of UCNPs at tumor site. Such enzyme-triggered UCNPs cross-linking would lead to the increased upconversion emission under $808\,\mathrm{nm}$ laser irradiation and subsequently amplified ROS generation, which thus

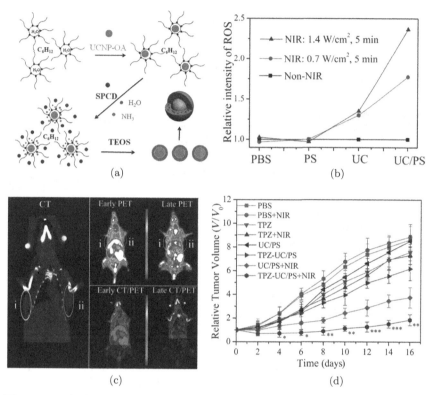

Fig. 5.4. a) Illustration of UCNPs coated with PS-doped dense silica formation process; b) ROS generation in cancer cells treated with PDT, as assessed by flow cytometry; c) CT, PET and CT/PET imaging of HeLa cell-bearing tumors pretreated with PS-loaded-UCNPs after intravenous injection of 18F-labeled MISO. i) tumor with NIR light (980 nm) irradiation, ii) tumor without any further treatment; d) The tumor growth curves of different groups after *in vivo* treatment. Reproduced with permission from Ref. 85. Copyright 2015, John Wiley & Sons, Inc.

greatly enhanced PDT treatment efficiency and multimodality cancer imaging including optical and PA imaging.

5.4. Surface Coated Carbon Nanotubes

Carbon nanotubes have attracted considerable attention in the past decades in many fields, from material science to biology applications.

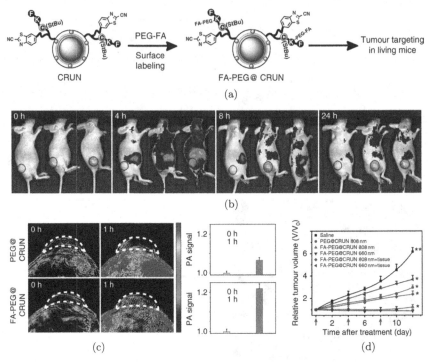

Fig. 5.5. a) Scheme illustration of the targeting strategy of FA-PEG@CRUN through the conjugation of FA and PEG on CRUN; b) Fluorescence imaging of tumours (blue circle) in living mice at different time intervals after injection (from left to right: Saline, PEG@CRUN, FA-PEG@CRUN, respectively; c) Photoacoustic imaging in the tumor site at two different time intervals (0 h and 1 h) after intravenous injection with PEG@CRUN (top) and FA-PEG@CRUN (middle), the right figure is the photoacoustic imaging signals; d) Tumour volumes change as a function of time in treated groups to evaluate the effectiveness of PDT treatment *in vivo*. Reproduced with permission from Ref. 92. Copyright 2016, Nature Publishing Group.

Typically, CNTs often own favorable sizes which made them working as ideal nanocarriers to deliver therapeutic agents to the disease site.[93] Similar to gold nanomaterials, CNTs have been extensively studied in biomedical sciences, mostly due to their intrinsic strong NIR optical absorbance (700–1000 nm).[17,94–97] In this context, we will mainly discuss CNT nanomaterials for NIR-controlled drug delivery, PTT and PDT.

5.4.1. *Drug delivery*

In recent years, there has been an increasing focus on the development of CNT nanoplatforms in biomedicine.[98,99] Qin and co-workers reported a Dox-delivery system through the amphiphilic biopolymer coated CNTs.[100] In this work, the CNTs were firstly functionalized with biopolymer CS through the hydrophobic alkyl chains, followed by further encapsulated into thermal sensitive pNIPAAm polymers. When the Dox-loaded polymer coated CNTs incubated with Hela cells for different time periods (e.g. 15 min, 2 h and 24 h *etc.*), followed by NIR laser irradiation ($\lambda = 808$ nm, 1 W/cm^2), the temperature induced by the hybrid CNT materials was increased, which subsequently initiated the Dox release. Therefore, the fluorescence intensity of Dox was found to apparently inhanced, which clearly indicated that NIR light excitation could effectively trigger the release of drug molecules from CNTs.

5.4.2. *Photothermal therapy*

One dimensional CNTs have also been commonly utilized as PTT agents because of their strong light absorbance in NIR region, which can cause in situ photothermal effect for NIR light-induced photothermal therapy.[93] As early as in 2005, a milestone reported by Dai and co-workers has utilized DNA-conjugated PEGylated CNTs for *in vitro* PTT applications.[101] This as-synthesized single-walled CNTs showed high photothermal efficiency to tumor cells and can induce remarkable cell death due to the thermal effect upon 2 min of 808 nm laser irradiation (with the power density of 1.4 W/cm^2). Furthermore, the PEGylated CNTs were functionalized with folate acid (FA) for selectively target the cancer cells, and thus allowing specific destruction of tumor cells through NIR laser irradiation. In continuation of the effective PTT treatment based on CNTs, many research breakthroughs have been constructed for further relevant applications in photothermal therapy. Moon *et al.* developed a biocompatible PEGlyated single walled CNTs for NIR-irradiated photothermal therapy *in vivo*.[102] Besides single-walled CNTs, in another work done by Zhang *et al.*, they demonstrated a novel strategy for simultaneous assemble of peptide-modified multi-walled CNTs in the presence of NIR light.[103] Therefore, more targeting multi-walled CNTs could be attracted in the tumor site and finally achieving effective photothermal cancer treatment. Amounts of research works have been performed to study CNTs for their photothermal applications *in vitro* and *in vivo*.[104,105]

5.4.3. *Photodynamic therapy*

Besides PTT application, these carbon-based nanomaterials could also serve as effective nanocarriers to deliver photosensitizers and chemotherapeutic molecules, and could thus promote their unique applications in PDT and chemotherapy. For example, CNTs have been used for loading photosensitizer molecules through non-covalent interaction, such as zinc phthalocyanine (ZnPc),[106] chlorinee6 (Ce6).[107] Chao *et al.* recently also combined luminescent Ru(II) polypyridyl complexes with SWCNT together.[105] Such Ru(II) complexes could be released from SWCNT based on the photothermal effect, and meanwhile produce $^1O_2$ upon the NIR laser irradiation (at 808 nm). The Hela cells which treated with Ru@SWCNTs and 2,7-dichlorodihydro fluorescein diacetate (DCFH-DA), would lead to significant fluorescence enhancement under 808 nm light irradiation ($0.25 \, W/cm^2$, 2 min), which suggesting the indicator of DCFH-DA can be oxidized by the generated $^1O_2$ in living cells.[108] For the *in vivo* examination, the Hela tumor model bearing nude mice were treated with Ru@SWCNTs samples through tail vein injection for 15 days, and followed by NIR laser irradiation (808 nm, $0.25 \, W/cm^2$, 5 min), which would result in the significant tumors ablation in living animals.

5.5. Surface Coated 2D Nanomaterials

Apart from the gold-related nanostructures, UCNPs and CNTs for NIR light-triggered imaging and therapy, a variety of surface coated 2D nanomaterials, including graphene oxides, molybdenum disulfide (MoS_2), black phosphorus (BP), have also been demonstrated for their promising NIR-responsive properties for NIR light-controlled therapies.

5.5.1. *Graphene oxide (GO)*

2D graphene nanomaterials have emerged as novel nanomaterials in NIR-responsive biomedicine, because of high absorption in the NIR window, good biocompatibility and large surface area. Zhong *et al.* fabricated Dox-loaded PEGylated nano-graphene oxide (Dox-PEG-GO) for the chemotherapy in 2011.[109] In this work, drug molecules Dox were loaded on the surface of nanographene oxide through physical absorption. The *in vivo* antitumor treatment was conducted by NIR light illumination ($2\,W/cm^2$) for 5 min, the Dox-PEG-GO nanocarriers treated tumor was found to be completely destructed without obvious weight loss. Besides, Matteini *et al.* also reported NIR light-controlled drug release systems by employing graphene oxide.[110] In this system, graphene and its derivatives have been known to play the dual roles either as a temporary storage for drug delivery, or as an effective transducer to convert NIR light into heat, which thus induces the drug release for promising chemotherapy. Additionally, in a work reported by Liu and co-workers, PEGylated GO labelled with fluorescent dye was first investigated *in vivo* photothermal treatment.[111] The nanocomplex showed high tumor passive uptake in three different tumor models, 4T1, KB and U87MG tumor bearing nude mice. They demonstrated that GO is an efficient photothermal agent which can generate heat under 808 nm laser irradiation (e.g. $2\,W/cm^2$) and exhibit effective tumor growth inhibition. Furthermore, GO can also be non-covalently functionalized with PS moelcules for PDT application.[112,113] For example, Zhang *et al.* have developed a nanocomposite by combine the UCNPs with graphene oxide through covalently bond in 2013.[114] The ZnPc photosensizer was loaded on the nanocomposite surface. This as-synthesized UCNP-GO/ZnPc theranostic nanoplatform could be used for the combinatorial PDT and PTT treatment for cancer therapy.

5.5.2. *Molybdenum disulfide*

Recently, a new type of two dimensional nanomaterials have been emerged in nanomedicine field partially because of their large surface area, which have been an ideal template to adsorb a large quantity of drug molecules. In 2011, the single-layer MoS_2 was first synthesized by chemical exfoliation.[115,116] Zhao's group have demonstrated a simple and low cost approach for NIR light photothermal-triggered drug delivery by using this single-layer MoS_2.[117] In their work, the commercially available MoS_2 were treated with oleum and sonicated to form a grayish dispersion. In order to increase the water solubility of MoS_2 nanosheets, CS biopolymer was introduced to coat the MoS_2 nanosheets. The commonly used anticancer drug Dox was loaded onto the MoS_2-CS complex by simply mixing with the MoS_2 solution (Fig. 5.6). KB cells were incubated with MoS_2-CS-Dox for 2 hours and followed by NIR light irradiation, red fluorescence signals inside cells were found to be greatly

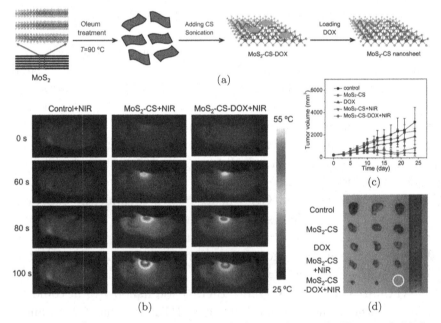

Fig. 5.6. a) Schematic illustration of the synthesis of CS coated MoS_2 nanosheets as a NIR photothermal-triggered drug delivery nanoplatform; b) Infrared thermal images of tumor bearing mice injected with saline as control, MoS_2-CS with NIR, and MoS_2-CS-DOX with NIR laser; c) Tumor growth curves of tumors after various treatments for five groups; d) Photograph of tumors from the control group, MoS_2-CS group, DOX group, MoS_2-CS +NIR group, and MoS_2-CS-DOX+NIR group. Reproduced with permission from Ref. 117. Copyright 2014, American Chemical Society.

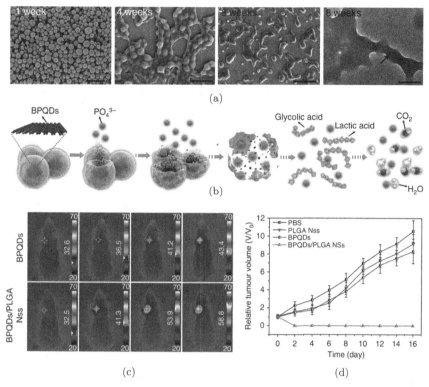

Fig. 5.7. a) SEM images of the BPQDs/PLGA NSs after degradation in PBS for 1, 4 and 8 weeks together with the corresponding TEM image of the nanospheres after degradation for 8 weeks; b) Schematic representation of the degradation process of the BPQDs/PLGA nanospheres in the physiological environment; c) Infrared thermal images of the MCF7 breast tumour-bearing nude mice irradiated by the 808 nm laser (1 W/cm²) at 24 h after separate intravenous injection with 100 ml of BPQDs and BPQDs/PLGA nanospheres; d) Growth curves of MCF7 breast tumor in different groups of nude mice treated with PBS, PLGA NSs, BPQDs and BPQDs/PLGA NSs with the NIR laser irradiation. Reproduced with permission from Ref. 120. Copyright 2016, Nature Publishing Group.

enhanced, indicating the drug molecules were released from the MoS_2 nanosheets. Furthermore, MoS_2-CS-Dox nanocomposites have been extensively carried out for *in vivo* experiment, the results as showed in the work clearly reveal that the inhibition of the tumor growth after NIR laser treatment.

5.5.3. *Black phosphorus*

BP is a new member of 2D nanomaterials, which expected to be an attractive theranostic agent mostly due to its biocompatibility and can degrade to nontoxic phosphate and phosphonate in aqueous media.[118,119] Shao and co-workers designed a biodegradable NIR-responsive photothermal ablation cancer treatment strategy by using BP-based nanospheres.[120] The BP quantum dots (~3 nm) were incorporated in PLGA polymer to form the BPQDs/PLGA nano-spheres for prolonging the blood circulation time in body. Such composite nanomaterials exhibit higher NIR absorbance, which were used for PTT (Fig. 5.7). As the result of the EPR effect, BPQDs/PLGA showed remarkably enhanced tumor accumulation property, thus photothermal ablation with excellent therapeutic outcomes was achieved in their animal experiments.

5.6. Conclusion and Current Challenges

In this chapter (format), the latest advances in NIR light-controlled therapies based on surface-coated nanomaterials have been discussed. Herein, we mainly concentrated on the surface coated gold-based nanostructures, surface coated UCNPs, surface coated CNTs and 2D nanomaterials, and focus on their three types of therapeutic applications: NIR light-mediated drug delivery, NIR-light controlled photothermal therapy and photodynamic therapy. Apart from their attractive biocompatibility as well as good physical and chemical properties, there are several other reasons attributing to their wide applications, of which, are also the major concerns in biomedical research.

Several kinds of nanomaterials, like gold-based nanostructures and black phosphorus, are generally considered to be safe for living beings because of its stable chemistry with appropriate sizes, surface functionalization and biodegradable. However, some other nanomaterials, especially those with inorganic components, are not biodegradable and could potentially retain inside the body for long periods after treatment process. Although numerous studies have broadly demonstrated that, with appropriate surface coatings, there is no noticeable toxicity observed in both *in vitro* and *in vivo* studies, the nanomaterials for biomedical applications are still under the cautious examinations whether the long-term safety can be achieved. In general, thanks to the rapid development of nanotechnologies and tremendous efforts made to this area, the final approval of the promising theranostic applications in clinics based on nanostructures can be predictable in near future.

The goal of surface coated nanostructure-assisted therapies is to maximize the therapeutic activity and meanwhile minimize the side effects. Although many strategies have been demonstrated for nanomaterials with appropriate surface coatings, there is no noticeable toxicity observed, the nanomaterials for biomedical applications are still under the cautious examinations. In this regard, various approaches for design the surface coated nanostructures should be developed. Additionally, the intrinsic property of the nanostructures is always at the heart of the new concept which meets the various demands of therapies, including drug-loading efficiency, controlled drug release rate, targeting and localized accumulations of therapeutic agents at specific diseases areas.

NIR light is more biocompatibility not only due to its minimum photo damage to living tissues, but also because of its significantly enhanced penetration depth in living tissues. By understanding of human organism responding to different wavelength light, NIR light excitations are more preferred over those irradiations in UV or visible windows. However, for the light-sensitive agents, they usually resort to UV light (with higher energy), such as the chemical "uncaging" process where the bond breaking may occur. In this concern, UCNPs may become an ideal alternative which help to overcome the current obstacle. Therefore, despite the success of UCNPs in principle, one drawback is that the upconversion process has relatively low quantum yield. Although many kinds of UCNPs have been made with improved converting efficiency by involving in various doped components, great efforts are still required to develop UCNPs that can be more effective to emit UV light when upon NIR light irradiation.

References

1. Willmann, J. K., van Bruggen, N., Dinkelborg, L. M. and Gambhir, S. S. (2008) *Nat. Rev. Drug Discov.* **7**, 591–607.
2. Koo, H., Huh, M. S., Sun, I. C., Yuk, S. H., Choi, K., Kim, K. and Kwon, I. C. (2011) *Acc. Chem. Res.* **44**, 1018–1028.

3. Wang, S., Teng, Z., Huang, P., Liu, D., Liu, Y., Tian, Y., Sun, J., Li, Y., Ju, H., Chen, X. and Lu, G. (2015) *Small* **11**, 1801–1810.

4. Bansal, A. and Zhang, Y. (2014) *Acc. Chem. Res.* **47**, 3052–3060.

5. Yang, Y. M., Aw, J. X., Chen, K., Liu, F., Padmanabhan, P., Hou, Y. L., Cheng, Z. and Xing, B. G. (2011). *Chem. Asian J.* **6**, 1381–1389.

6. Shao, Q. and Xing, B. (2012) *Chem. Commun.* **48**, 1739–1741.

7. Pelaz, B., Alexiou, C., Alvarez-Puebla, R. A., Alves, F., Andrews, A. M., Ashraf, S., Balogh, L. P., Ballerini, L., Bestetti, A., Brendel, C., Bosi, S., Carril, M., Chan, W. C. W., Chen, C., Chen, X., Chen, X., Cheng, Z., Cui, D., Du, J., Dullin, C., Escudero, A., Feliu, N., Gao, M., George, M., Gogotsi, Y., Grünweller, A., Gu, Z., Halas, N. J., Hampp, N., Hartmann, R. K., Hersam, M. C., Hunziker, P., Jian, J., Jiang, X., Jungebluth, P., Kadhiresan, P., Kataoka, K., Khademhosseini, A., Kopeček, J., Kotov, N. A., Krug, H. F., Lee, D. S., Lehr, C.-M., Leong, K. W., Liang, X.-J., Ling Lim, M., Liz-Marzán, L. M., Ma, X., Macchiarini, P., Meng, H., Möhwald, H., Mulvaney, P., Nel, A. E., Nie, S., Nordlander, P., Okano, T., Oliveira, J., Park, T. H., Penner, R. M., Prato, M., Puntes, V., Rotello, V. M., Samarakoon, A., Schaak, R. E., Shen, Y., Sjöqvist, S., Skirtach, A. G., Soliman, M. G., Stevens, M. M., Sung, H.-W., Tang, B. Z., Tietze, R., Udugama, B. N., VanEpps, J. S., Weil, T., Weiss, P. S., Willner, I., Wu, Y., Yang, L., Yue, Z., Zhang, Q., Zhang, Q., Zhang, X.-E., Zhao, Y., Zhou X. and Parak, W. J. (2017). *ACS Nano*.

8. Yang, Y., Velmurugan, B., Liu X. and Xing, B. (2013). *Small* **9**, 2937–2944.

9. Karimi, M., Sahandi Zangabad, P., Baghaee-Ravari, S., Ghazadeh, M., Mirshekari, H. and Hamblin, M. R. (2017). *J. Am. Chem. Soc.*

10. Yuan, Y., Zhang, C. J. and Liu, B. (2015). *Angew. Chem. Int. Ed.* **54**, 11419–11423.

11. Li, X., Mu, J., Liu, F., Tan, E. W. P., Khezri, B., Webster, R. D., Yeow, E. K. L. and Xing, B. (2015). *Bioconjugate Chem.* **26**, 955–961.

12. Yang, Y., Aw, J. and Xing, B. (2017). *Nanoscale* **9**, 3698–3718.

13. Heilman, B. J., St. John, J., Oliver, S. R. J. and Mascharak, P. K. (2012). *J. Am. Chem. Soc.* **134**, 11573–11582.

14. Chung, J. W., Lee, K., Neikirk, C., Nelson, C. M. and Priestley, R. D. (2012). *Small* **8**, 1693–1700.

15. Shao, Q. and Xing, B. (2010). *Chem. Soc. Rev.* **39**, 2835–2846.

16. Gnanasammandhan, M. K., Idris, N. M., Bansal, A., Huang, K. and Zhang, Y. (2016). *Nat. Protocols* **11**, 688–713.

17. Kim, H., Chung, K., Lee, S., Kim, D. H. and Lee, H. (2016). *WIREs Nanomed Nanobiotechnol* **8**, 23–45.

18. Yao, C., Wang, P., Li, X., Hu, X., Hou, J., Wang, L. and Zhang, F. (2016). *Adv. Mater.* **28**, 9341–9348.

19. Shanmugam, V., Selvakumar, S. and Yeh, C. S. (2014). *Chem. Soc. Rev.* **43**, 6254–6287.

20. Noh, M. S., Lee, S., Kang, H., Yang, J. K., Lee, H., Hwang, D., Lee, J. W., Jeong, S., Jang, Y., Jun, B. H., Jeong, D. H., Kim, S. K., Lee, Y. S. and Cho, M. H. (2015). *Biomaterials* **45**, 81–92.

21. Lv, S.-W., Liu, Y., Xie, M., Wang, J., Yan, X.-W., Li, Z., Dong, W.-G. and Huang, W.-H. (2016). *ACS Nano* **10**, 6201–6210.
22. Yang, Y., Mu, J. and Xing, B. (2017). *WIREs Nanomed Nanobiotechnol* **9**, n/a-n/a.
23. Yang, G., Gong, H., Liu, T., Sun, X., Cheng, L. and Liu, Z. (2015). *Biomaterials* **60**, 62–71.
24. Yang, G., Lv, R., He, F., Qu, F., Gai, S., Du, S., Wei Z. and Yang, P. (2015). *Nanoscale* **7**, 13747–13758.
25. Lv, R., Yang, P., He, F., Gai, S., Yang, G., Dai, Y., Hou, Z. and Lin, J. (2015). *Biomaterials* **63**, 115–127.
26. Cheng, L., Wang, C., Feng, L., Yang, K. and Liu, Z. (2014). *Chem. Rev.* **114**, 10869–10939.
27. Hu, Y., Yang, Y., Wang, H. and Du, H. (2015). *ACS Nano* **9**, 8744–8754.
28. Min, Y., Li, J., Liu, F., Padmanabhan, P., Yeow, E. and Xing, B. (2014). *Nanomaterials* **4**, 129.
29. Wang, S., Zhao, X., Wang, S., Qian J. and He, S. (2016). *ACS Appl. Mat. Interfaces* **8**, 24368–24384.
30. Lu, W., Melancon, M. P., Xiong, C., Huang, Q., Elliott, A., Song, S., Zhang, R., Flores, L. G., Gelovani, J. G., Wang, L. V., Ku, G., Stafford, R. J. and Li, C. (2011). *Cancer Res.* **71**, 6116–6121.
31. Li, J., Liu, J. and Chen, C. (2017). *ACS Nano*.
32. Zhang, Z., Wang, L., Wang, J., Jiang, X., Li, X., Hu, Z., Ji, Y., Wu, X. and Chen, C. (2012). *Adv. Mater.* **24**, 1418–1423.
33. Yang, X. J., Liu, X., Liu, Z., Pu, F., Ren, J. S. and Qu, X. G. (2012). *Adv. Mater.* **24**, 2890–2895.
34. Chang, Y. T., Liao, P. Y., Sheu, H. S., Tseng, Y. J., Cheng, F. Y. and Yeh, C. S. (2012). *Adv. Mater.* **24**, 3309–3314.
35. Zhang, Z., Wang, J., Nie, X., Wen, T., Ji, Y., Wu, X., Zhao, Y. and Chen, C. (2014). *J. Am. Chem. Soc.* **136**, 7317–7326.
36. Jeong, E. H., Ryu, J. H., Jeong, H., Jang, B., Lee, H. Y., Hong, S., Lee, H. and Lee, H. (2014). *Chem. Commun.* **50**, 13388–13390.
37. Liu, X., Gao, C., Gu, J., Jiang, Y., Yang, X., Li, S., Gao, W., An, T., Duan, H., Fu, J., Wang, Y. and Yang, X. (2016). *ACS Appl. Mat. Interfaces* **8**, 27622–27631.
38. You, J., Zhang, G. and Li, C. (2010). *ACS Nano* **4**, 1033–1041.
39. Zhong, Y., Wang, C., Cheng, L., Meng, F., Zhong, Z. and Liu, Z. (2013). *Biomacromolecules* **14**, 2411–2419.
40. Cong, H.-P., Qiu, J.-H. and Yu, S.-H. (2015). *Small* **11**, 1165–1170.
41. Rengan, A. K., Bukhari, A. B., Pradhan, A., Malhotra, R., Banerjee, R., Srivastava, R. and De, A. (2015). *Nano Lett.* **15**, 842–848.
42. Li, Q., Tang, Q., Zhang, P., Wang, Z., Zhao, T., Zhou, J., Li, H., Ding, Q., Li, W., Hu, F., Du, Y., Yuan, H., Chen, S., Gao, J., Zhan, J. and You, J. (2015). *Biomaterials* **57**, 1–11.
43. Lei, M., Ma, M., Pang, X., Tan, F. and Li, N. (2015). *Nanoscale* **7**, 15999–16011.
44. Delcea, M., Sternberg, N., Yashchenok, A. M., Georgieva, R., Bäumler, H., Möhwald, H. and Skirtach, A. G. (2012). *ACS Nano* **6**, 4169–4180.

45. Yavuz, M. S., Cheng, Y., Chen, J., Cobley, C. M., Zhang, Q., Rycenga, M., Xie, J., Kim, C., Song, K. H., Schwartz, A. G., Wang, L. V. and Xia, Y. (2009). *Nat. Mater.* **8**, 935–939.

46. Pitsillides, C. M., Joe, E. K., Wei, X. B., Anderson, R. R. and Lin, C. P. (2003). *Biophys. J.* **84**, 4023–4032.

47. Wu, Z., Lin, X., Wu, Y., Si, T., Sun, J. and He, Q. (2014). *ACS Nano* **8**, 6097–6105.

48. Hatef, A. and Meunier, M. (2015). *Opt. Express* **23**, 1967–1980.

49. Chen, W.-H., Luo, G.-F., Lei, Q., Hong, S., Qiu, W.-X., Liu, L.-H., Cheng, S.-X. and Zhang, X.-Z. (2017). *ACS Nano* **11**, 1419–1431.

50. Huang, P., Lin, J., Li, W., Rong, P., Wang, Z., Wang, S., Wang, X., Sun, X., Aronova, M., Niu, G., Leapman, R. D., Nie, Z. and Chen, X. (2013). *Angew. Chem. Int. Ed.* **52**, 13958–13964.

51. Su, Y., Wei, X., Peng, F., Zhong, Y., Lu, Y., Su, S., Xu, T., Lee, S.-T. and He, Y. (2012). *Nano Lett.* **12**, 1845–1850.

52. Zhang, L., Su, H., Cai, J., Cheng, D., Ma, Y., Zhang, J., Zhou, C., Liu, S., Shi, H. Zhang, Y. and Zhang, C. (2016). *ACS Nano* **10**, 10404–10417.

53. Jana, N. R., Gearheart, L. and Murphy, C. J. (2001). *J. Phys. Chem. B* **105**, 4065–4067.

54. Huang, X. H., El-Sayed, I. H., Qian, W. and El-Sayed, M. A. (2006). *J. Am. Chem. Soc.* **128**, 2115–2120.

55. von Maltzahn, G., Park, J.-H., Agrawal, A., Bandaru, N. K., Das, S. K., Sailor, M. J. and Bhatia, S. N. (2009). *Cancer Res.* **69**, 3892–3900.

56. Du, Y., Jiang, Q., Beziere, N., Song, L., Zhang, Q., Peng, D., Chi, C., Yang, X., Guo, H., Diot, G., Ntziachristos, V., Ding B. and Tian, J. (2016). *Adv. Mater.* **28**, 10000–10007.

57. Hou, Z., Deng, K., Li, C., Deng, X., Lian, H., Cheng, Z., Jin, D. and Lin, J. (2016). *Biomaterials* **101**, 32–46.

58. Cheng, Y., Doane, T. L., Chuang, C.-H., Ziady, A. and Burda, C. (2014). *Small* **10**, 1799–1804.

59. Vankayala, R., Huang, Y.-K., Kalluru, P., Chiang, C.-S. and Hwang, K. C. (2014). *Small* **10**, 1612–1622.

60. Yang, D., Ma, P. a., Hou, Z., Cheng, Z., Li, C. and Lin, J. (2015). *Chem. Soc. Rev.* **44**, 1416–1448.

61. Zhou, J., Liu, Q., Feng, W., Sun, Y. and Li, F. (2015). *Chem. Rev.* **115**, 395–465.

62. Wang, F., Han, Y., Lim, C. S., Lu, Y., Wang, J., Xu, J., Chen, H., Zhang, C., Hong, M. and Liu, X. (2010). *Nature* **463**, 1061–1065.

63. Jalani, G., Naccache, R., Rosenzweig, D. H., Haglund, L., Vetrone, F. and Cerruti, M. (2016). *J. Am. Chem. Soc.* **138**, 1078–1083.

64. Chen, G., Qiu, H., Prasad, P. N. and Chen, X. (2014). *Chem. Rev.* **114**, 5161–5214.

65. Yang, Y. M., Shao, Q., Deng, R. R., Wang, C., Teng, X., Cheng, K., Cheng, Z., Huang, L., Liu, Z., Liu, X. G. and Xing, B. G. (2012). *Angew. Chem. Int. Ed.* **51**, 3125–3129.

66. Fedoryshin, L. L., Tavares, A. J., Petryayeva, E., Doughan, S. and Krull, U. J. (2014). *ACS Appl. Mat. Interfaces* **6**, 13600–13606.

67. Min, Y., Li, J., Liu, F., Yeow, E. K. L. and Xing, B. (2014). *Angew. Chem. Int. Ed.* **53**, 1012–1016.
68. Dai, Y., Xiao, H., Liu, J., Yuan, Q., Ma, P. a., Yang, D., Li, C., Cheng, Z., Hou, Z. Yang, P. and Lin, J. (2013). *J. Am. Chem. Soc.* **135**, 18920–18929.
69. Jin, Q., Mitschang, F. and Agarwal, S. (2011). *Biomacromolecules* **12**, 3684–3691.
70. Jiang, J., Tong, X. and Zhao, Y. (2005). *J. Am. Chem. Soc.* **127**, 8290–8291.
71. Jayakumar, M. K. G., Idris N. M. and Zhang, Y. (2012). *PNAS* **109**, 8483–8488.
72. Lin, M., Gao, Y., Hornicek, F., Xu, F., Lu, T. J., Amiji, M. and Duan, Z. (2015). *Adv. Colloid Interface Sci.* **226, Part B**, 123–137.
73. Liu, Y., Kobayashi, T., Iizuka, M., Tanaka, T., Sotokawa, I., Shimoyama, A. Murayama, Y., Otsuji, E., Ogura, S.-i. and Yuasa, H. (2013). *Bioorg. Med. Chem.* **21**, 2832–2842.
74. Dong, B., Xu, S., Sun, J., Bi, S., Li, D., Bai, X., Wang, Y., Wang, L. and Song, H. (2011). *J. Mater. Chem.* **21**, 6193–6200.
75. Challenor, M., Gong, P., Lorenser, D., Fitzgerald, M., Dunlop, S., Sampson, D. D. and Swaminathan Iyer, K. (2013). *ACS Appl. Mat. Interfaces* **5**, 7875–7880.
76. Xiao, Q., Zheng, X., Bu, W., Ge, W., Zhang, S., Chen, F., Xing, H., Ren, Q., Fan, W., Zhao, K., Hua, Y. and Shi, J. (2013). *J. Am. Chem. Soc.* **135**, 13041–13048.
77. Mitsunaga, M., Ogawa, M., Kosaka, N., Rosenblum, L. T., Choyke, P. L. and Kobayashi, H. (2011). *Nat. Med.* **17**, 1685–1691.
78. Zou, W., Visser, C., Maduro, J. A., Pshenichnikov, M. S. and Hummelen, J. C. (2012). *Nat Photon* **6**, 560–564.
79. Ai, F., Ju, Q., Zhang, X., Chen, X., Wang, F. and Zhu, G. (2015). *Sci. Rep.* **5**, 10785.
80. Chen, Q., Wang, C., Cheng, L., He, W., Cheng, Z. and Liu, Z. (2014). *Biomaterials* **35**, 2915–2923.
81. Zhang, P., Steelant, W., Kumar, M. and Scholfield, M. (2007). *J. Am. Chem. Soc.* **129**, 4526–4527.
82. Wang, M., Chen, Z., Zheng, W., Zhu, H., Lu, S., Ma, E., Tu, D., Zhou, S., Huang, M. and Chen, X. (2014). *Nanoscale* **6**, 8274–8282.
83. Idris, N. M., Gnanasammandhan, M. K., Zhang, J., Ho, P. C., Mahendran, R. and Zhang, Y. (2012). *Nat. Med.* **18**, 1580–1585.
84. Cui, S., Yin, D., Chen, Y., Di, Y., Chen, H., Ma, Y., Achilefu, S. and Gu, Y. (2013). *ACS Nano* **7**, 676–688.
85. Liu, Y., Liu, Y., Bu, W., Cheng, C., Zuo, C., Xiao, Q., Sun, Y., Ni, D., Zhang, C., Liu, J. and Shi, J. (2015). *Angew. Chem. Int. Ed.* **54**, 8105–8109.
86. Lu, S., Tu, D., Hu, P., Xu, J., Li, R., Wang, M., Chen, Z., Huang, M. and Chen, X. (2015). *Angew. Chem. Int. Ed.* **54**, 7915–7919.
87. Fan, W., Bu, W. and Shi, J. (2016). *Adv. Mater.* **28**, 3987–4011.
88. Li, Y., Tang, J., Pan, D.-X., Sun, L.-D., Chen, C., Liu, Y., Wang, Y.-F., Shi, S. and Yan, C.-H. (2016). *ACS Nano* **10**, 2766–2773.
89. Wang, Y. F., Liu, G. Y., Sun, L. D., Xiao, J. W., Zhou, J. C. and Yan, C. H. (2013). *ACS Nano* **7**, 7200–7206.

90. Shen, J., Chen, G., Vu, A. M., Fan, W., Bilsel, O. S., Chang, C. C. and Han, G. (2013). *Adv. Optical Mater.* **1**, 644–650.

91. Xie, X. and Liu, X. (2012). *Nat. Mater.* **11**, 842–843.

92. Ai, X., Ho, C. J. H., Aw, J., Attia, A. B. E., Mu, J., Wang, Y., Wang, X., Wang, Y., Liu, X., Chen, H., Gao, M., Chen, X., Yeow, E. K. L., Liu, G., Olivo, M. and Xing, B. (2016). *Nat Commun* **7**, 10432.

93. Hong, G., Diao, S., Antaris, A. L. and Dai, H. (2015). *Chem. Rev.* **115**, 10816–10906.

94. Wong, B. S., Yoong, S. L., Jagusiak, A., Panczyk, T., Ho, H. K., Ang, W. H. and Pastorin, G. (2013). *Adv. Drug Deliver. Rev.* **65**, 1964–2015.

95. Hashida, Y., Tanaka, H., Zhou, S., Kawakami, S., Yamashita, F., Murakami, T., Umeyama, T., Imahori, H. and Hashida, M. (2014). *J. Controlled Release* **173**, 59–66.

96. Igarashi, T., Kawai, H., Yanagi, K., Cuong, N. T., Okada, S. and Pichler, T. (2015). *Phys. Rev. Lett.* **114**, 176807.

97. Hartland, G. V. (2011). *Chem. Rev.* **111**, 3858–3887.

98. Robinson, J. T., Hong, G., Liang, Y., Zhang, B., Yaghi, O. K. and Dai, H. (2012). *J. Am. Chem. Soc.* **134**, 10664–10669.

99. Liang, C., Diao, S., Wang, C., Gong, H., Liu, T., Hong, G., Shi, X., Dai, H. and Liu, Z. (2014). *Adv. Mater.* **26**, 5646–5652.

100. Qin, Y., Chen, J., Bi, Y., Xu, X., Zhou, H., Gao, J., Hu, Y., Zhao, Y. and Chai, Z. (2015). *Acta Biomater.* **17**, 201–209.

101. Kam, N. W. S., O'Connell, M., Wisdom, J. A. and Dai, H. (2005). *Proc. Nat. Acad. Sci. U.S.A.* **102**, 11600–11605.

102. Moon, H. K., Lee, S. H. and Choi, H. C. (2009). *ACS Nano* **3**, 3707–3713.

103. Zhang, B., Wang, H., Shen, S., She, X., Shi, W., Chen, J., Zhang, Q., Hu, Y., Pang, Z. and Jiang, X. (2016). *Biomaterials* **79**, 46–55.

104. Murakami, T., Nakatsuji, H., Inada, M., Matoba, Y., Umeyama, T., Tsujimoto, M., Isoda, S., Hashida, M. and Imahori, H. (2012). *J. Am. Chem. Soc.* **134**, 17862–17865.

105. Zhang, P., Huang, H., Huang, J., Chen, H., Wang, J., Qiu, K., Zhao, D., Ji, L. and Chao, H. (2015). *ACS Appl. Mat. Interfaces* **7**, 23278–23290.

106. Zhang, M., Murakami, T., Ajima, K., Tsuchida, K., Sandanayaka, A. S. D., Ito, O., Iijima, S. and Yudasaka, M. (2008). *PNAS* **105**, 14773–14778.

107. Zhu, Z., Tang, Z., Phillips, J. A., Yang, R., Wang, H. and Tan, W. (2008). *J. Am. Chem. Soc.* **130**, 10856–10857.

108. Pierroz, V., Joshi, T., Leonidova, A., Mari, C., Schur, J., Ott, I., Spiccia, L., Ferrari, S. and Gasser, G. (2012). *J. Am. Chem. Soc.* **134**, 20376–20387.

109. Zhang, W., Guo, Z., Huang, D., Liu, Z., Guo, X. and Zhong, H. (2011). *Biomaterials* **32**, 8555–8561.

110. Matteini, P., Tatini, F., Cavigli, L., Ottaviano, S., Ghini, G. and Pini, R. (2014). *Nanoscale* **6**, 7947–7953.

111. Yang, K., Zhang, S., Zhang, G., Sun, X., Lee, S.-T. and Liu, Z. (2010). *Nano Lett.* **10**, 3318–3323.

112. Ge, J., Lan, M., Zhou, B., Liu, W., Guo, L., Wang, H., Jia, Q., Niu, G., Huang, X., Zhou, H., Meng, X., Wang, P., Lee, C.-S., Zhang, W. and Han, X. (2014). *Nat Commun* **5**.

113. Sahu, A., Choi, W. I., Lee, J. H. and Tae, G. (2013). *Biomaterials* **34**, 6239–6248.

114. Wang, Y., Wang, H., Liu, D., Song, S., Wang, X. and Zhang, H. (2013). *Biomaterials* **34**, 7715–7724.

115. Lee, J., Kim, J. and Kim, W. J. (2016). *Chem. Mater.* **28**, 6417–6424.

116. Coleman, J. N., Lotya, M., O'Neill, A., Bergin, S. D., King, P. J., Khan, U., Young, K., Gaucher, A., De, S. and Smith, R. J. (2011). *Science* **331**, 568–571.

117. Yin, W., Yan, L., Yu, J., Tian, G., Zhou, L., Zheng, X., Zhang, X., Yong, Y. Li, J., Gu, Z. and Zhao, Y. (2014). *ACS Nano* **8**, 6922–6933.

118. Joshua, O. I., Gary, A. S., Herre, S. J. v. d. Z. and Andres, C.-G. (2015). *2D Materials* **2**, 011002.

119. Ling, X., Wang, H., Huang, S., Xia, F. and Dresselhaus, M. S. (2015). *PNAS* **112**, 4523–4530.

120. Shao, J., Xie, H., Huang, H., Li, Z., Sun, Z., Xu, Y., Xiao, Q., Yu, X.-F., Zhao, Y., Zhang, H., Wang, H. and Chu, P. K. (2016). *Nature Communications* **7**, 12967.

CHAPTER 6

Surface Functionalization through Polymer Segregation

TATSUO MARUYAMA*

Department of Chemical Science and Engineering,
Kobe University, Japan

6.1. Introduction

Polymers play a major role in daily commodities, as well as in biological and medical applications, because they are versatile, light and easily processed. The surface properties of solid materials determine their adsorption and immobilization performances, their textures, slipperiness and wettability, and their suitability as adhesives, coatings and paints. Many reports show that the morphology and composition of the surface of a polymeric material are different from those of the bulk phase. In general, one or more components (or segments) in the surface region of a solid polymeric material is spontaneously and preferentially enriched relative to the bulk composition; this is termed "surface segregation". In this chapter, surface-segregation approaches to the control of polymer surface properties for biological and environmental applications are reviewed and discussed.

Hydrophobic domains (or segments) of polymers are likely to be enriched at an air/polymer interface under dry conditions, which

---

*Corresponding author: tmarutcm@crystal.kobe-u.ac.jp

means that the surfaces of solid polymeric materials are relatively hydrophobic compared with their bulk phases. This phenomenon (surface segregation) is explained by minimization of the surface free energy at the air/polymer interface. In particular, fluorinated compounds, especially perfluoroalkyl (R_f) groups, are known to segregate at air/polymer interfaces (Fig. 6.1) because fluorinated surfaces have a remarkably low surface free energy.[1–8] Utilizing the unique properties of fluorinated compounds, Ye *et al.* reported that surface segregation of perfluoroalkyl groups in a random copolymer produced a surface that was resistant to protein adsorption (anti-fouling property).[8] They synthesized random copolymers of 2-perfluorooctylethyl methacrylate and 2-hydroxyethyl methacrylate with various monomer compositions (Fig. 6.2), and prepared the polymer film by spin-coating. X-ray photoelectron spectroscopy (XPS) and quartz crystal microbalance measurements revealed that the surface composition of the monomers was related to the mount of protein adsorption.

Ober *et al.* synthesized comb-like block copolymers and studied the adhesion of biofilms to the copolymer surfaces. Using two

Fig. 6.1. Schematic illustration of the surface segregation of perfluoroalkyl (R_f) groups on a polymer-coated surface.

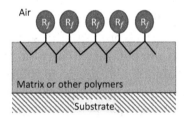

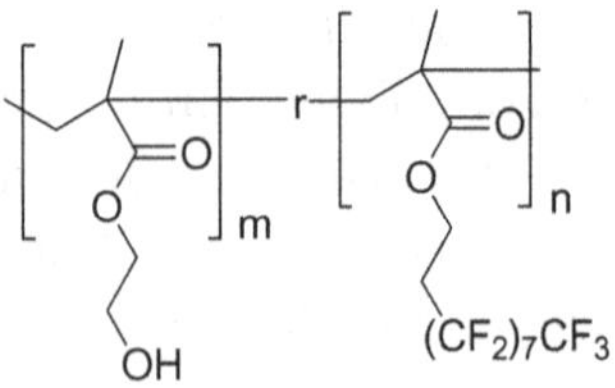

Fig. 6.2. Random copolymer containing perfluoroalkyl groups synthesized by Ye *et al.*[8]

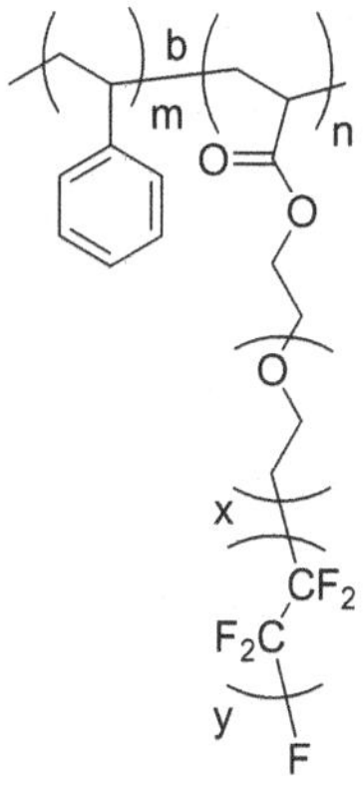

Fig. 6.3. Comb-like block copolymer synthesized by Ober *et al.*[9]

different algal species as model biofoulants — one adhered strongly
to hydrophobic surfaces and the other to hydrophilic surfaces — they
succeeded in reducing bacterial surface-fouling (biofouling)[9] when
the polymer side chains were composed of a perfluoroalkyl group
and an oligo(ethylene glycol) group (Fig. 6.3). The surface of the
copolymer film prepared by spin-coating exhibited weak adhesion to
both algal species. Surface segregation of the perfluoroalkyl group
was accompanied by surface segregation of the oligo(ethylene glycol)
group, which reduced the adhesion of biofilms.

Anti-fouling properties, or the resistance toward protein and
microbe adsorption (or adhesion), are also of great importance in
membrane filtration. Mayes and coworkers reported pioneering work
on the surface segregation of polymers and the preparation of
low-fouling surfaces.[10] For this, they used an interesting approach,
namely the entropically driven surface segregation of poly(ethylene
oxide) (PEG) chains. PEG is a surface modifier known for its
remarkable ability to resist protein adsorption; this is a result of
its hydrophilicity, large excluded volume and unique coordination
with the surrounding water molecules in aqueous solution.[11–13] The
authors demonstrated that simple blending of linear PEG with poly-
methylmethacrylate (PMMA) resulted in segregation of PEG away
from the surface, whereas a PEG-branched PMMA copolymer (containing a high content of PEG, insoluble in water) segregated to

the surface, leading to hydration of the surface. They attributed the enrichment of PEG moieties at the surface to the entropic effect of the PEG chains in the polymer. The PEG-displayed surface showed high resistance to protein adsorption and animal cell adhesion.

Mayes and coworkers also employed the surface-segregation strategy and phase-inversion method to prepare a porous poly(vinylidene fluoride) (PVDF) membrane, and successfully enriched the hydrophilic moieties on a surface to obtain a membrane with low-fouling properties.[14] To coagulate the PVDF (and thus fabricate a porous membrane), they immersed a PVDF solution containing an amphiphilic comb-like copolymer of methyl methacrylate and polyoxyethylene methacrylate in water. The aqueous environment induced the surface segregation of the hydrophilic polyoxyethylene group in the amphiphilic copolymer on the PVDF membrane surface, and this was revealed by XPS measurements. The addition of only 5 wt% of the amphiphilic copolymer to a PVDF membrane enhanced its resistance to protein fouling. The membrane surface also exhibited a self-healing property in water, which indicated that reconstruction of the polymer surface (segregation) occurred in water after polymer coagulation.

Nishino *et al.* reported a polymeric surface covered with perfluoroalkyl groups and PEG moieties, which was prepared using the surface segregation strategy.[15,16] They synthesized a methacrylate terpolymer containing perfluoroalkyl and PEG chains (Fig. 6.4), and prepared the terpolymer-segregated surface by dip-coating a solution mixture of the terpolymer, PMMA and vinylidene fluoride-tetrafluoroethylene copolymer. They found that both the perfluoroalkyl groups and the PEG moieties were segregated on the surface. Interestingly, the surface exhibited high water repellency in air and

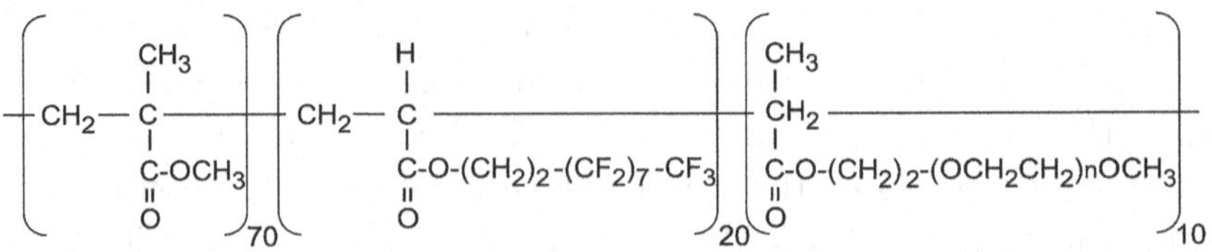

Fig. 6.4. Methacrylate terpolymer containing perfluoroalkyl chains and poly(ethylene oxide) chains synthesized by Nishino *et al.*[15,16]

high hydrophilicity in water. The surface also showed resistance to protein adsorption, resistance to antithrombogenicity and resistance to cell adhesion. Moreover, the authors also succeeded in the covalent immobilization of a protein on the low-fouling surface by utilizing the hydroxy groups at the termini of the surface PEG chains as reactive sites.

The incorporation of reactive functional groups on a polymeric surface is important for the immobilization of functional molecules. We have proposed a method for displaying amino groups (a reactive group) on a surface through polymer segregation.[17] We synthesized methacrylate copolymers containing amino groups with various protecting groups (Fig. 6.5) and evaluated the display of the amino groups on dip-coated copolymer surfaces. Protection of the amino groups affected their surface segregation, and *tert*-butoxycarbonyl (Boc) protection effectively displayed amino groups (more than $70 \, \text{pmol/cm}^2$) on the surfaces of a variety of polymer substrates, probably as a result of the high hydrophobicity of the Boc group. The Boc protection was easily removed by acid treatment and the surface amino groups were readily available for subsequent reactions (e.g. covalent immobilization).

To control the segregation of functional groups in a polymer, we have also proposed the use of a surfactant (Fig. 6.6).[18] We anticipated that a low-molecular-weight surfactant would segregate to a polymeric surface when a polymer solution containing an appropriate

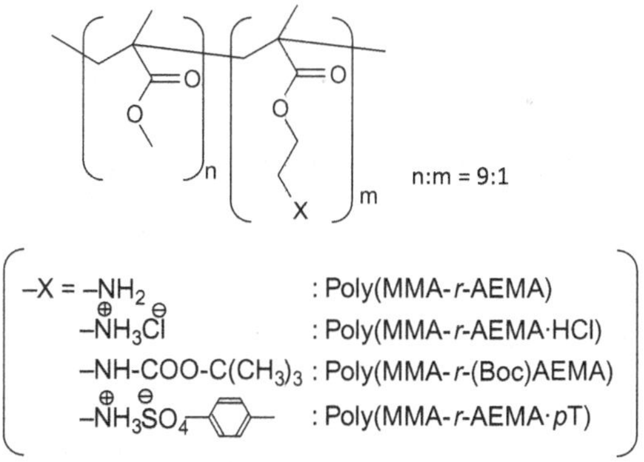

Fig. 6.5. Methacrylate copolymers containing amino groups with various protecting groups.[17]

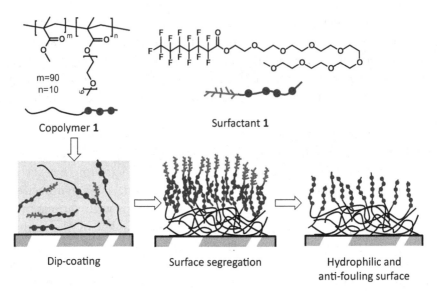

Fig. 6.6. Schematic illustration of surfactant-induced surface segregation of polymer side chains.[18]

surfactant was dip-coated and dried to form a polymer film in air. The surface segregation of the surfactant would also be accompanied by the segregation of functional groups in a polymer that interacts with the surfactant. We therefore synthesized a surfactant containing a hydrophilic oligo(ethylene glycol) group and a hydrophobic perfluoroalkyl group (surfactant 1, Fig. 6.6). Simple dip-coating of a solution of a methacrylate-based copolymer containing oligo(ethylene glycol) groups (copolymer 1) produced a hydrophobic surface because of surface segregation of the hydrophobic moiety in copolymer 1 during air-drying. In contrast, dip-coating of a mixed solution of copolymer 1 and surfactant 1 resulted in segregation of surfactant 1 on the top surface of the dip-coated layer owing to the high hydrophobicity of the perfluoroalkyl group. The oligo(ethylene glycol) groups in surfactant 1 were also involved in the segregation process, which meant that the oligo(ethylene glycol) groups of copolymer 1 were located just beneath the surface of the dip-coated layer. Surfactant 1 was easily removed from the surface simply by rinsing with water (copolymer 1 is insoluble in water), which afforded a surface with anti-fouling properties and composed of copolymer oligo(ethylene glycol) groups.

This approach can also be used to display reactive sites on a surface; carboxy groups were introduced at the termini of the oligo(ethylene glycol) groups of copolymer **1**, and the resulting dip-coated surface displayed surface carboxy groups that could be used for the immobilization of functional molecules.

In summary, a number of methods for the surface functionalization of solid polymeric materials have been reported, for example, plasma treatment, silane coupling treatment, graft polymerization, self-assembled monolayer formation, chemical vapor deposition and physical adsorption of an appropriate substance. Each method has its advantages and disadvantages. Although there are only a limited number of studies on surface functionalization using polymer segregation, surface segregation of polymers during the drying or annealing process is a practical and reliable method owing to its versatility and low cost. Interest and research in this area is expected to develop rapidly in a broad range of scientific fields, and this will greatly extend the knowledge of polymer surface chemistry and the practical utility of polymeric surfaces.

References

1. Thomas, R. R., Anton, D. R., Graham, W. F., Darmon, M. J., Sauer, B. B., Stika, K. M. and Swartzfager, D. G. (1997). Preparation and surface properties of acrylic polymers containing fluorinated monomers, *Macromolecules* **30**, 2883–2890.
2. Iyengar, D. R., Perutz, S. M., Dai, C. A., Ober, C. K. and Kramer, E. J. (1996). Surface segregation studies of fluorine-containing diblock copolymers, *Macromolecules* **29**, 1229–1234.
3. Jannasch, P. (1998). Surface structure and dynamics of block and graft copolymers having fluorinated poly(ethylene oxide) chain ends, *Macromolecules* **31**, 1341–1347.
4. Boker, A., Reihs, K., Wang, J. G., Stadler, R. and Ober, C. K. (2000). Selectively thermally cleavable fluorinated side chain block copolymers: Surface chemistry and surface properties, *Macromolecules* **33**, 1310–1320.
5. de Grampel, R. D. V., Ming, W., Gildenpfennig, A., van Gennip, W. J. H., Laven, J., Niemantsverdriet, J. W., Brongersma, H. H., de With, G. and van der Linde, R. (2004). The outermost atomic layer of thin films of fluorinated polymethacrylates, *Langmuir* **20**, 6344–6351.
6. Ming, W., Tian, M., van de Grampel, R. D., Melis, F., Jia, X., Loos, J. and van der Linde, R. (2002). Low surface energy polymeric films from solventless liquid oligoesters and partially fluorinated isocyanates, *Macromolecules* **35**, 6920–6929.

7. Xue, D. W., Wang, X. P., Ni, H. G., Zhang, W. and Xue, G. (2009). Surface segregation of fluorinated moieties on random copolymer films controlled by random-coil conformation of polymer chains in solution, *Langmuir* **25**, 2248–2257.

8. Zhao, Z. L., Ni, H. G., Han, Z. Y., Jiang, T. F., Xu, Y. J., Lu, X. L. and Ye, P. (2013). Effect of surface compositional heterogeneities and microphase segregation of fluorinated amphiphilic copolymers on antifouling performance, *ACS. Appl. Mater. Interfaces* **5**, 7808–7818.

9. Krishnan, S., Ayothi, R., Hexemer, A., Finlay, J. A., Sohn, K. E., Perry, R., Ober, C. K., Kramer, E. J., Callow, M. E., Callow, J. A. and Fischer, D. A. (2006). Anti-biofouling properties of comblike block copolymers with amphiphilic side chains, *Langmuir* **22**, 5075–5086.

10. Walton, D. G., Soo, P. P., Mayes, A. M., Allgor, S. J. S., Fujii, J. T., Griffith, L. G., Ankner, J. F., Kaiser, H., Johansson, J., Smith, G. D., Barker, J. G. and Satija, S. K. (1997). Creation of stable poly(ethylene oxide) surfaces on poly(methyl methacrylate) using blends of branched and linear polymers, *Macromolecules* **30**, 6947–6956.

11. Kjellander, R. and Florin, E. (1981). Water-structure and changes in thermal-stability of the system poly(ethylene oxide)-water, *Journal of the Chemical Society-Faraday Transactions I* **77**, 2053-+.

12. Lee, J. H., Lee, H. B. and Andrade, J. D. (1995). Blood compatibility of polyethylene oxide surfaces, *Prog. Polym. Sci.* **20**, 1043–1079.

13. Elbert, D. L. and Hubbell, J. A. (1996). Surface treatments of polymers for biocompatibility, *Annu. Rev. Mater. Sci.* **26**, 365–394.

14. Hester, J. F., Banerjee, P. and Mayes, A. M. (1999). Preparation of protein-resistant surfaces on poly(vinylidene fluoride) membranes via surface segregation, *Macromolecules* **32**, 1643–1650.

15. Tokuda, K., Kawasaki, M., Kotera, M. and Nishino, T. (2015). Highly water repellent but highly adhesive surface with segregation of poly(ethylene oxide) side chains, *Langmuir* **31**, 209–214.

16. Tokuda, K., Noda, M., Maruyama, T., Kotera, M. and Nishino, T. (2015). A Low-fouling polymer surface prepared by controlled segregation of poly(ethylene oxide) and its functionalization with biomolecules, *Polymer Journal* **47**, 328–333.

17. Shimomura, A., Nishino, T. and Maruyama, T. (2013). Display of amino groups on substrate surfaces by simple dip-coating of methacrylate-based polymers and its application to DNA immobilization, *Langmuir* **29**, 932–938.

18. Yamamoto, S., Kitahata, S., Shimomura, A., Tokuda, K., Nishino, T. and Maruyama, T. (2015). Surfactant-induced polymer segregation to produce antifouling aurfaces via dip-coating with an amphiphilic polymer, *Langmuir* **31**, 125–131.

CHAPTER 7

Nucleic Acid Hairpins: A Robust and Powerful Motif for Molecular Devices

SUDHANSHU GARG,[*] HIEU BUI,[*]
ABEER ESHRA,[*,‡] SHALIN SHAH[†]
and JOHN REIF[*,§]

[*]Department of Computer Science,
Duke University, Durham, NC 27708, USA
[†]Department of Electrical and Computer Engineering,
Duke University, Durham, NC 27708, USA
[‡]Department of Computer Science and Engineering,
Faculty of Electronic Engineering,
Menoufia University, Menouf, Menoufia 32952, Egypt

7.1. Introduction

DNA was not discovered in 1953. Instead, Watson and Crick famously discovered the *double helical structure* of DNA in 1953.[66] Neither was it discovered in 1944, when Oswald Avery showed that DNA was the carrier of hereditary information.[3] It was 75 years prior, in 1869, when Friedrich Miescher,[17] a young swiss doctor, while trying to determine the constituents of leukocytes, discovered

---

[§]Corresponding author: reif@cs.duke.edu

a material that originated from the nucleus (discovered in 1802[4]), and named it *nuclein*.[37] Mieschers original preparation failed to separate the protein protamine, and when Kossel, Richard Altmann and co-workers succeeded in identifying and removing this protein, they named it Nuclensure (nucleic acid),[1] believing it to be a subunit of *nuclein*, and since it had properties of an acid (a fact Miescher had already discovered earlier[17]). Kossel was responsible for identifying the constituents of *nuclein*, namely the purine and pyrimidine bases, one sugar and phosphoric acid. In addition, he discovered that *nuclein* was an important constituent of chromatin, and that this material is not a source of energy, but is related to the synthesis of protoplasm during growth and development,[30] for which he was awarded the Nobel Prize (1910). In the subsequent decades, interest in nucleic acids was lost, partly due to the tetranucleotide hypothesis advanced by Phoebus Levene and Hermann Steudel (reviewed in Olby[43]). This hypothesis was rebutted in 1951 by Refs. 13, 14, leading to the discovery of the double helical structure of nucleic acids and subsequent discovery of the genetic code.[a]

7.2. What are Hairpins?

7.2.1. *DNA hairpins*

DNA hairpins are a special *secondary structure* formed by an ssDNA, and contain a neck/stem double stranded region, and an unhybridized loop region, as seen in Fig. 7.1. Hairpins have been recognized as a useful tool in molecular computation because of 3 reasons: (1) Hairpins store energy in their unhybridized loop, and on hybridization, energy is released driving the reaction forward. (2) In their hairpin form, they are relatively unreactive with other DNA strands, and act as excellent monomers until an external entity (usually another DNA strand) causes the stem region to open and react with other DNA complexes. Hence, they can persist with low leaks for a long amount of time. (3) A common way to create DNA complexes

---

[a]This discussion has been adapted from an article by Ref. 17, which makes for a very exciting read.

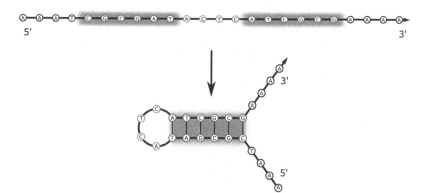

Fig. 7.1. Hairpin open and closed forms. The stem is highlighted and the loop consists of four nucleotides but it can be any length.

is to anneal them. DNA complexes usually contain a large number of strands, and multiple different structures can be formed because of varied interactions between different strands. In low concentrations, DNA hairpins usually form without error, and are not involved in spurious structure formation. This is because their formation is not diffusion dependent, i.e. the two ends of a hairpin hybridize with each other before two ends of different hairpins hybridize. This property, is known as *locality*, and is a strong motivation for the use of hairpins.

DNA hairpins,[12,18,22,51,52,55] and metastable DNA hairpin complexes[2,23,50] have been used as fuel in chain reactions to form large polymers,[10,47] in programming pathways in self-assembly,[46] and in logic circuits.[20,49] A common technique to help open a DNA hairpin is via a process known as *toehold mediated strand displacement*.[6]

DNA hairpin has been extensively used in the field of DNA nanotechnology and beyond, because it stores energy its loop region. Since this can drive the reaction forward, it becomes an interesting candidate as a fuel for autonomous DNA hybridization reactions.

In a recent study by Green *et al.*,[55] a quantitative picture of DNA hairpin system, as shown in Fig. 7.2(a), which can act as a fuel was presented. In particular, they made a parametric model of DNA hairpin and tuned these parameters to study their output dependence. As shown in Fig. 7.2(b), hairpin structure has 3 parameters: m, n,

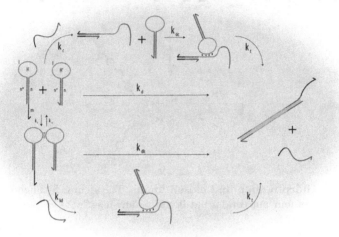

(a) **DNA hairpin interactions**

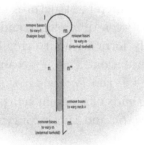

(b) **Modelling DNA hairpin**

Fig. 7.2. **DNA hairpin system as fuel for autonomous reactions.** (a) Illustration of all the possible interactions between two hairpins and external catalyst, including all the intermediate structures. (b) DNA hairpin as a model where length of toehold (m), neck (n), and loop (l) can be changed. The variation in the reaction rate with neck length (n) and loop length (l) is relatively less as compared to the length of toehold (m). Figure was reproduced with permission from Ref. 55.

and l; which can be represented as H(m, n, l). By tuning this model, they found following key properties of nucleic acid hairpins:

- For longer loops (l), they may formed spontaneous long-lived kissed complexes which can lead to substantial reduction in their role as fuel. This reaction leak can be reduced by increasing neck length (n) or decreasing loop length (l).

- Loop opening for a hairpin, by a catalyst, can remove steric hindrance which usually formed kissed complex. Therefore, a hairpin duplex is formed much faster in presence of catalyst.
- The toehold mediated strand displacement reaction of catalyst is 10–100 times faster with external toehold as compared to internal toehold.

Keeping these key points in mind, the complimentary hairpin system can be used a meta-stable fuel for autonomous DNA hybridization reactions.

7.2.2. *Kissed complex*

The formation of kissed complex is a spontaneous reaction which strong dependence of loop length (l). It was observed that kissed complexes were formed for loop length ranging from 14–40 but not shorter than that. Two small domains of length 6 on each side of neck also didn't form kissed complexes. Also, for longer loop length ($l = 40$), the kissed complex formed spontaneously and persisted for longer time intervals as shown in Fig. 7.3.

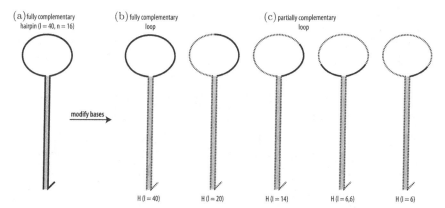

Fig. 7.3. The effect of loop length on kissed complex. Original hairpin with loop length 40 and neck length 16 was modified to observe kissing effect, which is the hybridization among hairpins within the loop region. Dark line indicate domains complimentary to other hairpin. For the hairpin with long loop, kissed complex not only formed spontaneously but also persisted for long time periods. Figure was reproduced with permission from Ref. 55.

7.2.3. *Catalytic opening strand*

A catalytic strand to open one of the hairpins was added in different amounts to study its effect on hairpins. Its concentration greatly reduced the kissed hairpin complex as well as hairpin monomers. However, the presence of intermediate complexes at high catalyst concentration was also observed.[55]

7.2.4. *Internal and external loop opening*

Ref. 55 tuned toehold length (m), neck length (n), loop length (l), and location of toehold. The catalytic strand which opens one of the hairpin loops is greatly influenced by the location of the toehold. In particular, the study identified a significant difference of one to two orders of magnitude in reaction rate between external and internal toehold. For shorter internal toehold, a decrease in loop length lead to a significant decrease in reaction rate. However, for external toehold, loop length is almost independent. There is a slight variation in the rates due to neck length and it is not monotonic as well; however, it is negligible as compared to the rate difference due to internal and external toehold.

7.2.5. *Interest in nucleic acid hairpins*

Nucleic acid hairpins (both RNA and DNA hairpins) have been a subject of study for the last four decades. Scientists stabilize single stranded RNA, serve as nucleation sites for RNA folding, protein recognition signals, mRNA localization and regulation of mRNA degradation.[64] On the other hand, DNA hairpins in biological contexts have been studied with respect to forming cruciform structures that can regulate gene expression. Inverted repeats often occur at potential recognition sites for proteins, operator sequences or transcription termination regions. In supercoiled DNA, inverted repeats result in cruciform structures, that can serve as recognition elements by certain proteins and thus have functional significance.[38,42] One example of utilizing a hairpin loop is the termination sequence for transcription in some prokaryotes. Once a polymerase encounters this loop, it falls off and transcription ends.

7.3. An Energy Perspective: How Much Power Do Hairpin Based Systems Really Pack?

The energy of nucleic acid hairpins lies in its single stranded region, that is usually any region other than the stem. On an average, each base pair on hybridization releases about $-1.4\,kcal/mol$[55] of energy. Assuming most reactions occur at a concentration of $100\,nM$, and in a volume of $100\,\mu L$, this would account for a net amount of $10\,pmol$ of reactants. If the number of bases per reactant is b, then the amount of energy released is $b \times -1.4\,kcal/mol \times 10\,pmol$ $\times$ reactants, which is $b \times -14\,ncal$. For the HCR system,[47] $12\,nt$ per reactant, and 2 reactants, thus, $12 \times 2 \times -14\,ncal$, or $336\,ncal$. To put this in perspective, one step that you walk consumes about $50\,mcal$. Thus walking a single step is equivalent to running 1.48×10^5 HCR systems.

7.4. Nucleic Acid Circuits

Hairpins have been utilized as a source of fuel for many different DNA circuits. Nucleic acid circuits with the ability to perform boolean logic operations have been increasingly been analyzed, with the goal of achieving point of care diagnostics. Using the looped region of DNA as a mechanism to store energy, and catalytically release it, was first described in 2003.[2] This was optimized in 2004 by the suggestion of opening *both* the looped regions using the same catalyst,[49] and was further optimized in its use as a fuel for a catalytic system.[23] This catalytic system formed an integral part of the logic circuits demonstrated later in 2006.[48] Other logic DNA circuits that use the bulge loop to store energy,[35,58] have been demonstrated.

DNA hairpins have also been used to construct a device capable of allosteric regulation.[69] As the construction of hairpin based signal gain devices improve, DNA hairpin systems are finding more adoption. More recently, DNA hairpins have been used in conjunction with enzyme based strand amplification mechanisms, SDA[59] and EXPAR[63] to construct two-input logic gates that detect RNA biomarkers (microRNAs as inputs).[7] The HCR system has been extended to 2D in order to increase the sensitivity of the system.[68] Improved logic gates such as AND & NAND[25] have been designed.

7.5. DNA Detection Mechanisms

DNA hairpins are commonly used as reactants in isothermal low cost DNA detection mechanisms. What makes them a material of choice, is their robustness and ability to sequester and release information as needed. They can be activated in three different ways, via hybridization: (1) External toehold[55] (2) Internal toehold[55] and (3) Molecular Beacon.[62] They have also been used in enzymatic systems with high sensitivity and specificity.

Molecular beacons are a very important practical way that DNA hairpins has been used. The first demonstration of this work involved a 25 *nt* hairpin loop with a 5 *nt* stem (Fig. 7.4). The ends of the strand are functionalized with a fluorophore and quencher molecule. In the presence of a target strand, the target on binding to the hairpin loop region, becomes a stiff rod (persistence length of DNA is 50 *nm* $\approx$ 150 *nt*), resulting in dehybridization of the stem, and a detectable change in fluorescence.

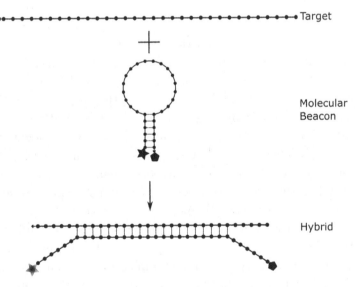

Fig. 7.4.   Operation of a molecular beacon. A nucleic acid hairpin is usually present in a closed state, with the fluorophore and quencher in close proximity (<10 nm). On the introduction of a target strand, the hairpin changes to an open state, resulting in an increase in fluorescence.

7.5.1. *Hybridization chain reaction*

Amplification is vital for bio-sensing applications like triggering nucleic acid aptamers. This can be accomplished by constructing biosensors from using DNA molecules. In particular, Ref. 47 first introduced the construction of hybridization chain reaction (HCR) for amplifying target molecules. In HCR, stable DNA species form a monomer only when exposed to an initiator fragment. In the absence of the initiator, the species coexist in solution with no interaction. The authors showed a simple novel device that consists of two hairpins (H1 and H2) and a single strand initiator (I) as illustrated in Fig. 7.5(a). Each hairpin has a sticky end at the end of one of the stem branches. *H*1's sticky end is a complement to a toehold in *I*. Strand displacement takes place when *I* is present in the solution. That opens the *H*1 revealing a previously sequestered toehold within the hairpin loop as seen in Fig. 7.5(b). The new toehold is complementary to *H*2's sticky end. A new strand displacement happens opening the *H*2 that has a hidden toehold that is complementary to the *H*1's sticky end as shown in Fig. 7.5(c). This reaction keeps going until the supply of hairpins in solution ends. The authors showed that the molecular weight of the resulting

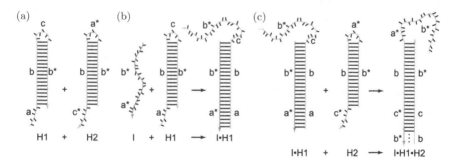

Fig. 7.5. Hybridization chain reaction by Ref. 47. (a) Hairpins H1 and H2 are stable in the absence of initiator I. (b) I nucleates at the sticky end of H1 and undergoes an unbiased strand displacement interaction to open the hairpin. (c) The newly exposed sticky end of H1 nucleates at the sticky end of H2 and opens the hairpin to expose a sticky end on H2 that is identical in sequence to I. Hence, each copy of I can propagate a chain reaction of hybridization events between alternating H1 and H2 hairpins to form a nicked double-helix, amplifying the signal of initiator binding. Figure was adopted with permission from Ref. 47.

products is reversely proportional to initiator concentration. Kinetic experiments and gel electrophoresis were performed to support the results. The strength of this system comes from storing energy within short loops that are protected with long stems. This puts hairpins in kinetic trap, preventing rapid equilibrium. Using DNA or RNA aptamers to create molecular recognition events have been demonstrated by the HCR system as well. The authors demonstrated that the aptamer opens the hairpin. This exposes the initiator strand of the HCR system. Authors also showed how to differentiate between ATP and GTP. Since HCR represents linear growth with initiator, it can be applied to generating more complicated sets of monomers like branched structures. Besides, result detection can be carried out with either gel electrophoresis or fluorescence detection. Both methods are cost effective. Therefore, HCR amplification can be used for both amplification and capturing target molecules. HCR has the characteristics of being protein-free and working room temperature. That makes it a good alternative to PCR.

A system with high specificity and sensitivity, that amplifies a few copies of DNA to as much as 10^9 in under an hour, is LAMP.[41] This system uses 6 primers initially, and then 4 primers in order to detect a DNA target strand. Although its primers itself are not hairpins, but its mechanism involves a system of inverted repeats in order to amplify the target, which is a demonstration of how stem-loops maybe used in detection.

DNA detection mechanisms need to amplify a target strand in order to bring them to levels that can be detected by traditional mechanisms such as fluorescence and gel electrophoresis. A robust mechanism for such amplification is Catalytic Hairpin Assembly (CHA), first demonstrated by Ref. 46, and then optimized by Ref. 33 as seen in Fig. 7.6. This mechanism has been coupled with two mechanisms, HCR[34] and LAMP[32] resulting in a higher level of sensitivity. In addition, multiple such circuits have been stacked together to achieve a million fold amplification.[15] Leaks in CHA have been improved tremendously,[27] paving the way for development of robust leak-free systems. Systems using these techniques have now been implemented inside live cells.[67]

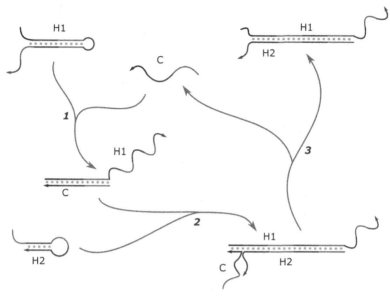

Fig. 7.6. Catalytic hairpin assembly by Ref. 33. (1) Hairpins H1 and H2 are stable in the absence of initiator C. (2) C nucleates at the sticky end of H1 and undergoes an unbiased strand displacement interaction to open the hairpin. (3) The newly exposed sticky end of H1 nucleates at the sticky end of H2 and opens the hairpin to expose a sticky end on H2 that is identical in sequence to C. This sticky end of H2 strand displaces C partially. The toehold of C still needs to dissociate, post which C is available to catalyze another pair of hairpins H1 & H2. Figure was reproduced with permission from Ref. 33.

7.6. Locomotive Devices, Fuel

DNA hairpins as fuel for molecular devices, including locomotion, were first demonstrated in 2003.[2] They were interesting because they bring to the table an *on-demand* energy supply mechanism.[11] The energy is hidden, until required. The energy is harnessed by opening the stem region, and exposing the loop. This is in contrast to a single stranded fuel with no secondary structure, where the energy is always available. The hidden energy coupled with programmability provides another functionality, of selectively choosing what reactions to hide and what reactions to allow to proceed, that helps develop a topological sequence of events.

DNA walkers have been built using one of two mechanisms:[54] (1) burnt-bridges[44] where the dsDNA is cut and need to reach forward to the next stator (foothold), (2) toehold mediated strand displacement[6] where additional DNA strands assist in detaching a walker from one stator and allowing it to find and bind to another stator. DNA hairpins have primarily been used in the second mechanism. One of the first autonomous DNA walkers using hairpins as fuel was performed using 4-way branch migration by Ref. 57. This mechanism was improved by Ref. 46, moving to 3-way branch migration, which significantly improved on the kinetics. In their scenario, there are 5 sites on a track. The cargo is initially bound to two sites, and in the presence of a hairpin fuel, site 1 is free to interact downstream with site 3 and so on. Also, this walker was stochastic, and could move in either direction, based on where the input hairpin fuel attached.

This design was succeeded in the same year, by an autonomous random catalytic hairpin design.[56] The core idea here is that the leading leg signals the back leg of the walker, to move forward, aided by DNA hairpins. Their design helps the walker move randomly in either direction. In 2009, this core idea was extended by Ref. 44, where the leading leg again signals the back leg of the walker, releasing it and moving it forward. However, in this case, the previous leg is "burnt", since the hairpin fuels bind to them irreversibly, and thus the back leg is forced to move unidirectionally.

Figure 7.7 shows the use of DNA hairpins as fuel in a programmable landscape. Here, instructions are *encoded* as part of the hairpin fuel strand, and as part of the leading leg. Cargo is moved on a DNA track, based on the input set of fuel hairpins. For one set of fuel hairpins, the cargo is programmed to move from X to Y. For another set of fuel hairpins, the cargo can be moved from X to Z.

7.7. An Enzyme-free Autocatalytic Self-replicating DNA Nanodevice

DNA hairpins have also been used in the operation of autocatalytic DNA nanodevices. It has been conjectured that enzyme-free self-replicating systems composed entirely of nucleic acids that operate in an autocatalytic fashion may have been the origin of life. In particular, Ref. 21 presents a novel enzyme-free autocatalytic self-replicating

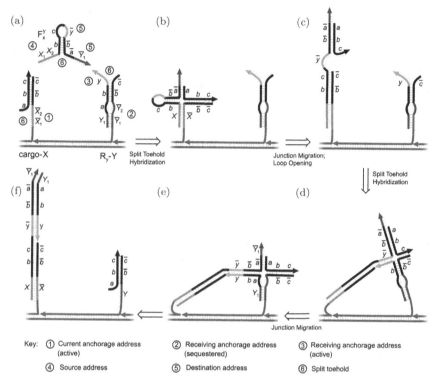

Fig. 7.7. A programmable molecular robot. DNA hairpins are used as fuel complexes to help in cargo movement from stator X to stator Y. In its initial state as a hairpin, only a toehold that can bind the hairpin to stator X is available. The toehold that can bind the hairpin to stator Y is hidden in the hairpin loop. Post branch migration, this toehold is revealed, and the cargo can be moved. Figure was adopted with permission from Ref. 39.

system composed entirely of DNA molecules that operates isothermally.[21] The non-enzymatic self-replicating system is constructed using DNA hairpins and DNA duplexes which self-assemble into a 3-arm junction upon activation by an initiator sequence. DNA hairpin motifs are used as the source of fuel, and the DNA duplexes serve as transducers to enable communication between two different replicating systems. A pair of transducers can be used to make this system cross-catalytic,[21] and therefore autocatalytic and capable of producing self-assembled 3-arm junctions, the number of which is exponentially growing in time. Triggered self-replicating systems can be used as autocatalytic isothermal DNA strand detector systems.[21] Leaks are a major challenge in designing robust DNA based detectors, affecting specificity and sensitivity of these systems, and to minimize leaks we developed a number of design techniques to avoid self-replication before reception of the initiator sequence. These included highly optimized sequence design, purification of DNA strands, and most notably a novel technique (irreversible hairpin opening) that both reduces leaks and improves on the kinetics of these systems.

7.7.1. *Transducers*

A transducer is a set of two dsDNA complexes as seen in Fig. 7.8(b) & e. The transducer helps insulate Replicator A sequences from Replicator B. Using the leakless DNA strand displacement designs that are a two-layer linear cascade[61] involving duplexes, it helps reducing the leak caused solely by the transducer. The term TA is used to encapsulate the two duplexes TA-Fuel1 and TA-Fuel2, and likewise TB for duplexes TB-Fuel1 and TB-Fuel2. Of the two duplexes TA-Fuel1 and TA-Fuel2, the top strand of TA-Fuel2 acts as an initiator for Replicator B. Similarly, the top strand of TB-Fuel2 acts as an initiator for Replicator A. Together, the two systems work cross-catalytically, and result in exponential kinetics. Note that the entire process is enzyme free, and takes place via toehold-mediated-strand-displacement.[6]

The concentration of transducers TA and TB can help control the rate of the reaction. The initial state of the system consists of 6 metastable hairpins and 4 duplexes. The hairpins have been designed using clamps[46] in order to prevent leaks. An initiator strand (for system A) at low concentration is introduced into this system. Replicator A is formed, and it starts the catalytic process. At reaction saturation, we expect to see Replicators A and B completely formed. The lower the initiator strand concentration that can work without leaks sets the sensitivity of the system. Figure 7.9 gives the complete mechanism of the self-replication that have mentioned previously.

3-arm Junction & Transducer Functioning: An important design consideration in the 3-arm junction, is that in Replicator A, hairpin A3 does not completely dissociate the initiator strand from

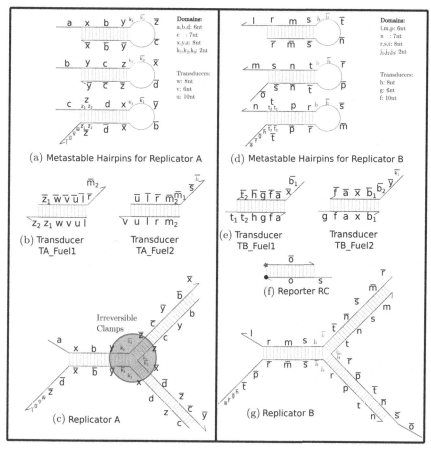

Fig. 7.8. Components of autocatalytic system. All the components of replicators A (left) and B (right) are shown. (a) & (d) The 3 metastable hairpins for Replicator A (B), in their initial conformations, (b) & (e) Transducer TA (TB), consists of two duplexes TA_Fuel1 and TA_Fuel2,[61] (c) & (g) The structures of replicators A & B, and the irreversible clamps in the middle of the junction, that ensure each hairpin opening step is irreversible. Also, the reporting is done via Replicator B, by the strand $\bar{o}\bar{s}$, that opens up a reporter complex (f) with a fluorophore and quencher.

Replicator A, i.e. the TB-Fuel2-Top strand, but it allows the initiator strand to float away, i.e. the domains a and d* are orthogonal and do not bind to each other. Likewise in Replicator B. This design aspect was introduced by Ref. 33, and helps to reduce leaks in the system.[21]

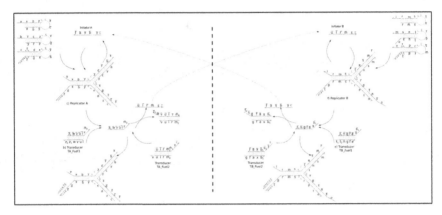

Fig. 7.9. Complete cross-catalytic mechanism. Replicator A interacting with Transducer TA to release catalyst for initiating formation of Replicator B. Similar system for Replicator B. Replicator B also binds to a reporter complex and releases the fluorophore.

7.7.2. *Irreversible hairpin opening*

Traditional DNA hairpin opening involves an invading ssDNA that binds to the hairpin via an external (or internal) toehold, and opens up the stem region of the hairpin. One factor that has not been discussed is that this reaction is reversible. This is because the incumbent arm of the hairpin, is still in local proximity to the invading strand, and can re-displace the invading strand, and the toehold region floats away (this is also the basis of remote toehold-mediated strand displacement). The incumbent arm has a *permanent toehold*, since it is covalently bound to the hairpin.[21] has experimentally verified this behavior via PAGE electrophoresis.

For the proposed reactions to be biased forward, the hairpin opening is needed to be irreversible. This can be engineered if the invading strand has a small clamp domain after the stem region. In this case, the invading strand can't be re-displaced by the incumbent arm, since the incumbent arm is non-complementary to this clamp domain region, and thus cannot initiate strand displacement as illustrated in Fig. 7.10. This is crucial for two reasons: (1) an opened hairpin is now permanently biased to react downstream, and is not present in intermediate branch migration states, (2) the kinetic rate of the hairpin opening reaction increases, since there is no reverse

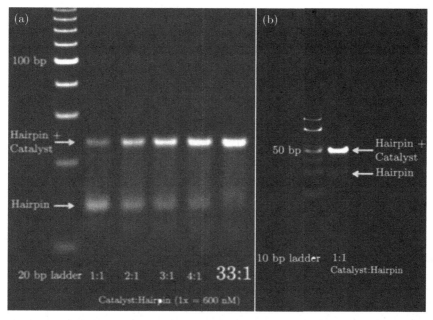

Fig. 7.10. Traditional and irreversible hairpin openings using external toehold. (a) A single hairpin without clamp domains does not completely open at 1:1 catalyst:hairpin concentration. Infact only when the reaction is biased by addition of excess catalyst, does the hairpin open. (b) A single hairpin with a clamp domain is completely opened at 1:1 catalyst:hairpin concentration.

reaction. Thus, there aren't fluctuations between the open and close hairpin states as was common earlier.

Another mechanism to bias the reaction forward, and make it irreversible, is to increase the length of the toehold. This mechanism was not chosen, because even though the thermodynamics bias the reaction forward, the incumbent arm of the hairpin is not always available for downstream reaction, and is constantly fluctuating between bound and unbound states, which affects the kinetics of the reaction. Two lengths of clamp domains (i.e. 2 nt and 4 nt) have been tested and the system works well in both scenarios.[21] It is possible that the system will work well with a 1 nt clamp domain as well, but this has not been investigated, since it has been shown that strand displacement can proceed with a single domain mismatch.[36] It is also important to note, that the clamp method described above, can be replaced by another method, in particular, if one can ensure that the incumbent arm has a region that forms a secondary structure, then the incumbent arm is not free to re-displace the invading strand. However, the secondary structure shouldn't be so strong that it cannot interact with a downstream strand. The toehold region that has to interact with the downstream strand should be left single stranded if this design method is pursued. One system that this design directly affects is HCR where the need for sequence optimization and sequence landscape investigation is crucial.[47] With this clamp domain, the reactions will always be biased forward, and the need for extensive optimization of sequence designs can be relaxed.

7.8. Summary

This chapter has demonstrated a multitude of molecular devices built from hairpin motifs. This has been possible due to studies that elucidate the thermodynamic and kinetic properties[8,31,53,65] of these motifs, leading to more predictable and robust devices. More recently, DNA hairpins have been used as a mechanism to help in assembly of DNA tetrahedrons,[5] including motifs such as TectoRNA.[19] The kissing complex plays an important role in genomic HIV-1 RNA[45] and as dynamic functional motifs.[9] With nucleic acid hairpins having such a diverse set of roles to play, they will be the subject of many molecular devices in the years to come.

7.9. Future Challenges

7.9.1. *Future extensions to self-replication and detection in vivo*

Enzymatic amplification systems have achieved the goal of high sensitivity and specificity, and can perform single molecule detection (Table 7.1). However, they do suffer from certain drawbacks. (i) High cost: They need a specific cocktail of enzymes to work (polymerase, or a combination of polymerase and a restriction enzyme) and in the presence of other enzymes might behave unexpectedly. (ii) They cannot be used in-vivo: (a) non-isothermal mechanisms such as PCR

Table 7.1. Performance of various enzymatic detection mechanisms.

Ref.	Year	[Reactant]	[Target]/ Amplification fold/Signal-to-Background ratio	Type of system, reactant purity
[41]	2000	0.2–0.8 μM	6 copies/ **NIF/NIF**	Isothermal, BST Polymerase, 45 mins @ 65°C
[60]	1992	1 μM	10–100 copies/ **NIF/NIF**	Isothermal,
[59]	1992	1 μM	10–50 copies/ **NIF/NIF**	exo⁻ Klenow, HincII, 1-5h @ 37°C Isothermal,
[26]	1993	2 μM	100 copies/ **NIF/NIF**	exo⁻ Klenow, HincII, 1-5h @ 37°C Real Time PCR,
Commercial Vendors (Applied Biosystems, Bio-rad, etc.)	2015	2 μM	1 copy/ **NIF/NIF**	Taq, 30–45 mins, 6 log units dynamic range Real Time PCR, Taq, 30–45 mins, 10 log units dynamic range

cannot be used, (b) the cell cytoplasm is frequently occupied by a motley of enzymes that can interfere with the functioning of the system.

Similarly enzyme-free detection systems (Table 7.2) also have drawbacks. (i) Although they are projected to work in-vivo, they are vulnerable to nucleases in the cell (however, modifications to the sugar, base and backbone such as phosphorothioate inter-nucleotide linkages, 2'O-methyl ribose modifications can help protect against degradation).[16] (ii) They do not create covalent bonds on replication, and thus are formed from weak hydrogen bonds, that are again more vulnerable than their covalent bonded counterparts. (iii) Their basic unit is often a single strand, or a duplex, hairpin etc., while the basic unit in enzymatic systems is a nucleotide. Thus, synthesis via enzymatic systems can achieve a higher level of granularity. Future

Table 7.2. Performance of various catalytic/autocatalytic enzyme-free nanodevices. Here, D-pure, DN-pure and DNB-pure are taken from existing nomenclature.[15]

Ref.	Year	[Reactant]	[Target]/ Amplification fold/ Signal-to-Background ratio	Type of system, reactant purity
[46]	2008	20 nM	100 pM/200X/1.15	2-layer circular cascade, D-pure
[33]	2011	50 nM	200 pM/50X/2.0	1-layer catalytic system, D-pure
[15]	2013	100 nM	5 nM/9X/2.0	2-layer linear cascade, D-pure
		100 nM	5 nM/20X/12.0	2-layer linear cascade, DN-pure
		100 nM	20 pM/600X/1.5	2-layer linear cascade, DNB-pure
		100 nM	10 pM/7000X/4.0	2-layer linear cascade, DNB-pure
		50–200 nM	20 pM/600,000X/1.3	4-layer linear cascade, Enzymatic & DN-pure 2-Stage
[27]	2014	50 nM	2.5 nM/13X/>100.0	1-layer catalytic system, D-pure
[40]	2015	100 nM	**<1 nM/>42X/8.25**[*]	1-layer auto-catalytic system, D-pure

[*]Data inferred from article.

DNA detection systems need to be developed with protections[24] to provide reliable operation in-vivo and in the cell.

7.9.2. *Future extensions to self-replication of large DNA nanostructures*

Extending self-replication to larger nanostructures such as DNA origami or DNA bricks[28,29] is the next logical step, since it can enable increased sensitivity, easier readouts from cellular environments, and reduce the need for cellular uptake of nanostructures, where they can be synthesized and amplified directly inside the cell.

References

1. Altmann, Richard (1889). Ueber nucleinsäuren, *Arch. Anat. Physiol. Abt. Physiol* **1889**, 524–536.
2. Andrew Turberfield, James Mitchell, Bernard Yurke, Allen Mills, Myrtle Blakey and Friedrich Simmel (2003). DNA fuel for free-running nanomachines, *Physical Review Letters* **90**, 11.
3. Avery, Oswald T., MacLeod, Colin M. and McCarty, Maclyn (1944). Studies on the chemical nature of the substance inducing transformation of pneumococcal types induction of transformation by a desoxyribonucleic acid fraction isolated from pneumococcus type III, *The Journal of experimental medicine* **79**(2), 137–158.
4. Baker, John R (1948). The cell theory: A restatement, history and critique, *Quarterly Journal of Microscopical Science* **89**(1), 103–125.
5. Barth, Anna, Kobbe, Daniela and Focke, Manfred (2016). DNA–DNA kissing complexes as a new tool for the assembly of DNA nanostructures, *Nucleic Acids Research*, gkw014.
6. Bernard Yurke, Andrew Turberfield, Allen Mills, Friedrich Simmel and Jennifer Neumann (2000). A DNA-fuelled molecular machine made of DNA, *Nature* **406**(6796), 605–608.
7. Bi, Sai, Ye, Jiayan, Dong, Ying, Li, Haoting and Cao, Wei (2016). Target-triggered cascade recycling amplification for label-free detection of microRNA and molecular logic operations, *Chemical Communications* **52**(2), 402–405.
8. Bois, Justin S., Venkataraman, Suvir, Choi, Harry M. T., Spakowitz, Andrew J., Wang, Zhen-Gang and Pierce, Niles A. (2005). Topological constraints in nucleic acid hybridization kinetics, *Nucleic acids research* **33**(13), 4090–4095.
9. Brunel, Christine, Marquet, Roland, Romby, Pascale and Ehresmann, Chantal (2002). RNA loop–loop interactions as dynamic functional motifs, *Biochimie* **84**(9), 925–944.
10. Bui, H., Garg, S., Miao, V., Song, T., Mokhtar, R. and Reif, J. (2017a). Design and analysis of linear cascade dna hybridization chain reactions using dna hairpins, *New Journal of Physics* **19**(1), 015006, http://stacks.iop.org/1367-2630/19/i=1/a=015006.
11. Bui, H., Miao, V., Garg, S., Mokhtar, R., Song, T. and Reif, J. (2017b). Design and analysis of localized DNA hybridization chain reactions, *Small* pp. 1602983–n/adoi:10.1002/smll.201602983, http://dx.doi.org/10.1002/smll.201602983.
12. Bui, H., Shah, S., Mokhtar, R., Song, T., Garg, S. and Reif, J. (2018). Localized DNA hybridization chain reactions on DNA origami, *ACS Nano* **12**(2), 1146–1155.
13. Chargaff, Erwin (1951). Structure and function of nucleic acids as cell constituents. in *Federation proceedings*, Vol. 10, p. 654.
14. Chargaff, Erwin, Vischer, Ernst, Doniger, Ruth, Green, Charlotte and Misani, Fernanda (1949). The composition of the desoxypentose nucleic acids of thymus and spleen, *Journal of Biological Chemistry* **177**(1), 405–416.

15. Chen, Xi, Briggs, Neima, McLain, Jeremy R. and Ellington, Andrew D. (2013). Stacking nonenzymatic circuits for high signal gain, *Proceedings of the National Academy of Sciences* **110**(14), 5386–5391, doi:http://dx.doi.org/10. 1073/pnas.1222807110, http://www.pnas.org/content/110/14/5386.abstract.

16. Chen, Yuan-Jyue, Groves, Benjamin, Muscat, Richard A. and Seelig, Georg (2015). DNA nanotechnology from the test tube to the cell, *Nature nanotechnology* **10**(9), 748–760.

17. Dahm, Ralf (2008). Discovering DNA: Friedrich Miescher and the early years of nucleic acid research, *Human Genetics* **122**(6), 565–581.

18. Eshra, A., Shah, S., Song, T. and Reif, J. (2019). Renewable DNA hairpin-based logic circuits, *IEEE Transactions on Nanotechnology*, **18**, 252–259, doi: 10.1109/TNANO.2019.2896189.

19. Feldkamp, Udo and Niemeyer, Christof M. (2006). Rational design of DNA nanoarchitectures, *Angewandte Chemie International Edition* **45**(12), 1856–1876.

20. Fu, D., Shah, S., Song, T. and Reif J. (2018). DNA-Based Analog Computing. In *Synthetic Biology* (pp. 411–417). Humana Press, New York, NY.

21. Garg, S., Gopalkrishnan, N., Chandran, H. and Reif, J. (2017). An autocatalytic self replicating dna nanodevice.

22. Garg, S., Shah, S., Bui, H., Song, T., Mokhtar, R. and Reif, J. (2018). Renewable time-responsive DNA circuits, *Small* **14**(33), 1801470.

23. Georg Seelig, Bernard Yurke and Erik Winfree (2006). Catalyzed relaxation of a metastable DNA fuel, *Journal of the American Chemical Society* **128**(37), 12211–12220.

24. Goltry, Sara, Hallstrom, Natalya, Clark, Tyler, Kuang, Wan, Lee, Jeunghoon, Jorcyk, Cheryl, Knowlton, William B., Yurke, Bernard, Hughes, William L. and Graugnard, Elton (2015). DNA topology influences molecular machine lifetime in human serum, *Nanoscale* **7**(23), 10382–10390.

25. Guo, Yuehua, Wu, Jie and Ju, Huangxian (2015). Target-driven DNA association to initiate cyclic assembly of hairpins for biosensing and logic gate operation, *Chemical Science* **6**(7), 4318–4323.

26. Higuchi, Russell, Fockler, Carita, Dollinger, Gavin and Watson, Robert (1993). Kinetic PCR analysis: Real-time monitoring of DNA amplification reactions, *Biotechnology* **11**, 1026–1030.

27. Jiang, Yu Sherry, Bhadra, Sanchita, Li, Bingling and Ellington, Andrew D. (2014). Mismatches improve the performance of strand-displacement nucleic acid circuits, *Angewandte Chemie* **126**(7), 1876–1879, doi:http://dx.doi.org/ 10.1002/ange.201307418, http://dx.doi.org/10.1002/ange.201307418.

28. Ke, Yonggang, Ong, Luvena L., Shih, William M. and Yin, Peng (2012). Three-dimensional structures self-assembled from DNA bricks, *Science* **338**(6111), 1177–1183, doi:http://dx.doi.org/10.1126/science.1227268, http:// science.sciencemag.org/content/338/6111/1177.

29. Ke, Yonggang, Ong, Luvena L., Sun, Wei, Song, Jie, Dong, Mingdong, Shih, William M. and Yin, Peng (2014). DNA brick crystals with prescribed depths, *Nature Chemistry*.

30. Kossel, Albrecht (1913). *Beziehungen der Chemie zur Physiologie* (BG Teubner).

31. Kuznetsov, Serguei V., Shen, Yiqing, Benight, Albert S. and Ansari, Anjum (2001). A semiflexible polymer model applied to loop formation in DNA hairpins, *Biophysical Journal* **81**(5), 2864–2875.
32. Li, Bingling, Chen, Xi and Ellington, Andrew D. (2012). Adapting enzyme-free DNA circuits to the detection of loop-mediated isothermal amplification reactions, *Analytical Chemistry* **84**(19), 8371–8377.
33. Li, Bingling, Ellington, Andrew D. and Chen, Xi (2011). Rational, modular adaptation of enzyme-free DNA circuits to multiple detection methods, *Nucleic Acids Research* **39**(16), e110, doi:http://dx.doi.org/10.1093/nar/gkr504, http://nar.oxfordjournals.org/content/39/16/e110.abstract.
34. Li, Bingling, Jiang, Yu, Chen, Xi and Ellington, Andrew D. (2012). Probing spatial organization of DNA strands using enzyme-free hairpin assembly circuits, *Journal of the American Chemical Society* **134**(34), 13918–13921.
35. Liu, Wenbin, Shi, Xiaohong, Zhang, Shemin, Liu, Xiangrong and Xu, Jin (2004). A new DNA computing model for the NAND gate based on induced hairpin formation, *Biosystems* **77**(1), 87–92.
36. Machinek, Robert R. F., Ouldridge, Thomas E., Haley, Natalie E. C., Bath, Jonathan and Turberfield, Andrew J. (2014). Programmable energy landscapes for kinetic control of DNA strand displacement, *Nature Communications* **5**.
37. Miescher-Rüsch, Friedrich (1871). *Ueber die chemische Zusammensetzung der Eiterzellen.*
38. Mizuuchi, Kiyoshi, Mizuuchi, Michiyo and Gellert, Martin (1982). Cruciform structures in palindromic DNA are favored by DNA supercoiling, *Journal of Molecular Biology* **156**(2), 229–243.
39. Muscat, Richard A., Bath, Jonathan and Turberfield, Andrew J. (2011). A programmable molecular robot, *Nano Letters* **11**(3), 982–987, doi: http://dx.doi.org/10.1021/nl1037165, http://dx.doi.org/10.1021/nl1037165.
40. Niranjan Srinivas (2015). *Programming chemical kinetics: Engineering dynamic reaction networks with DNA strand displacement*, Ph.D. thesis, California Institute of Technology.
41. Notomi, Tsugunori, Okayama, Hiroto, Masubuchi, Harumi, Yonekawa, Toshihiro, Watanabe, Keiko, Amino, Nobuyuki and Hase, Tetsu (2000). Loop-mediated isothermal amplification of DNA, *Nucleic Acids Research* **28**(12), e63, doi:http://dx.doi.org/10.1093/nar/28.12.e63, http://nar.oxfordjournals.org/content/28/12/e63.abstract.
42. Okazaki, Tuneko and Okazaki, Reiji (1969). Mechanism of DNA chain growth, IV. Direction of synthesis of T4 short DNA chains as revealed by exonucleolytic degradation, *Proceedings of the National Academy of Sciences* **64**(4), 1242–1248.
43. Olby, R. and Olby, R. (1994). The path to the double helix: The discovery of DNA, revised edition.
44. Omabegho, Tosan, Sha, Ruojie and Seeman, Nadrian C. (2009). A bipedal DNA Brownian motor with coordinated legs, *Science* **324**(5923), 67–71.

45. Paillart, Jean-Christophe, Skripkin, Eugene, Ehresmann, Bernard, Ehresmann, Chantal and Marquet, Roland (1996). A loop-loop "kissing" complex is the essential part of the dimer linkage of genomic HIV-1 RNA, *Proceedings of the National Academy of Sciences* **93**(11), 5572–5577.

46. Peng Yin, Harry Choi, Colby Calvert and Niles Pierce (2008). Programming biomolecular self-assembly pathways, *Nature* **451**(7176), 318–322.

47. Robert Dirks and Niles Pierce (2004). Triggered amplification by hybridization chain reaction, *Proceedings of the National Academy of Sciences of the United States of America* **101**(43), 15275–15278.

48. Seelig, Georg, Soloveichik, David, Zhang, David Yu and Winfree, Erik (2006). Enzyme-free nucleic acid logic circuits, *Science* **314**(5805), 1585–1588.

49. Seelig, Georg, Yurke, Bernard and Winfree, Erik (2004). DNA hybridization catalysts and catalyst circuits, in *DNA*, Vol. 3384 (Springer), 329–343, http://dblp.uni-trier.de/db/conf/dna/dna2004.html#SeeligYW04.

50. Shah, S. and Reif, J. (2018). Temporal DNA Barcodes: A Time-Based Approach for Single-Molecule Imaging. In *International Conference on DNA Computing and Molecular Programming 2018 Oct 8* (pp. 71–86). Springer, Cham.

51. Shalin Shah, Abhishek K. Dubey and John Reif (2019). Improved optical multiplexing with temporal DNA barcodes, *ACS Synthetic Biology* **8**(5), 1100–1111 doi:10.1021/acssynbio.9b00010.

52. Shalin Shah, Abhishek K. Dubey and John Reif (2019). Programming temporal DNA barcodes for single-molecule fingerprinting, *Nano Letters* **19**(4), 2668–2673 doi:10.1021/acs.nanolett.9b00590.

53. Shen, Yiqing, Kuznetsov, Serguei V. and Ansari, Anjum (2001). Loop dependence of the dynamics of DNA hairpins, *The Journal of Physical Chemistry B* **105**(48), 12202–12211.

54. Sherman, William (2009). Building a better nano-biped, *Science* **324**(5923), 46–47.

55. Simon Green, Daniel Lubrich and Andrew Turberfield (2006). DNA hairpins: Fuel for autonomous DNA devices, *Biophysical Journal* **91**(8), 2966–2975.

56. Simon Green, Jonathan Bath and Andrew Turberfield (2008). Coordinated chemomechanical cycles: A mechanism for autonomous molecular motion, *Physical Review Letters* **101**, 23.

57. Suvir Venkataraman, Robert Dirks, Paul Rothemund, Erik Winfree and Niles Pierce (2007). An autonomous polymerization motor powered by DNA hybridization, *Nature Nanotechnology* **2**, 490–494.

58. Takahashi, Keiichiro, Yaegashi, Satsuki, Kameda, Atsushi and Hagiya, Masami (2005). Chain reaction systems based on loop dissociation of DNA, in *DNA Computing* (Springer), pp. 347–358.

59. Terrance Walker, Melinda Fraiser, James Schram, Michael Little, James Nadeau and Douglas Malinowski (1992). Strand displacement amplification — An isothermal, in vitro DNA amplification technique, *Nucleic Acid Research* **20**(7), 1691–1696.

60. Terrance Walker, Michael Little, James Nadeau and Daryl Shank (1992). Isothermal in vitro amplification of DNA by a restriction enzyme/DNA polymerase system, *Proceedings of the National Academy of Sciences of the United States of America* **89**(1), 392–396.

61. Thachuk, Chris, Winfree, Erik and Soloveichik, David (2015). Leakless DNA strand displacement systems, in *DNA Computing and Molecular Programming* (Springer), pp. 133–153.

62. Tyagi, Sanjay, Kramer, Fred Russell and others (1996). Molecular beacons: Probes that fluoresce upon hybridization, *Nature Biotechnology* **14**(3), 303–308.

63. Van Ness, Jeffrey, Van Ness, Lori K. and Galas, David J. (2003). Isothermal reactions for the amplification of oligonucleotides, *Proceedings of the National Academy of Sciences* **100**(8), 4504–4509, doi:http://dx.doi.org/10.1073/pnas.0730811100, http://www.pnas.org/content/100/8/4504.abstract.

64. Varani, Gabriele (1995). Exceptionally stable nucleic acid hairpins, *Annual Review of Biophysics and Biomolecular Structure* **24**(1), 379–404.

65. Wallace, Mark I., Ying, Liming, Balasubramanian, Shankar and Klenerman, David (2001). Non-Arrhenius kinetics for the loop closure of a DNA hairpin, *Proceedings of the National Academy of Sciences* **98**(10), 5584–5589.

66. Watson, James D., Crick, Francis H. C. and others (1953). Molecular structure of nucleic acids, *Nature* **171**(4356), 737–738.

67. Wu, Cuichen, Cansiz, Sena, Zhang, Liqin, Teng, I-Ting, Qiu, Liping, Li, Juan, Liu, Yuan, Zhou, Cuisong, Hu, Rong, Zhang, Tao and others (2015). A nonenzymatic hairpin DNA cascade reaction provides high signal gain of mRNA imaging inside live cells, *Journal of the American Chemical Society* **137**(15), 4900–4903.

68. Xu, Yao and Zheng, Zhi (2016). Direct RNA detection without nucleic acid purification and PCR: Combining sandwich hybridization with signal amplification based on branched hybridization chain reaction, *Biosensors and Bioelectronics* **79**, 593–599.

69. Zhang, David Yu and Winfree, Erik (2008). Dynamic allosteric control of noncovalent DNA catalysis reactions, *Journal of the American Chemical Society* **130**(42), 13921–13926.

Index

World Scientific Series in Nanoscience and Nanotechnology

(Continuation of series card page)

Yiding Liu (Southwest Petroleum University, China),
Le He (Soochow University, China), Yihan Zhu (Zhejiang University
of Technology, China) and Yu Han (King Abdullah University of
Science and Technology, Saudi Arabia)

Vol. 17 *World Scientific Reference of Hybrid Materials*
(In 3 Volumes)
Volume 1: Block Copolymers
Volume 2: Devices from Hybrid and Organic Materials
Volume 3: Sol-Gel Strategies for Hybrid Materials
editor-in-chief Mato Knez (CIC Nanoscience Research Center, Spain)

Vol. 16 *World Scientific Handbook of Metamaterials and Plasmonics*
(In 4 Volumes)
Volume 1: Electromagnetic Metamaterials
Volume 2: Elastic, Acoustic, and Seismic Metamaterials
Volume 3: Active Nanoplasmonics and Metamaterials
Volume 4: Recent Progress in the Field of Nanoplasmonics
edited by Stefan A Maier (Imperial College London, UK)

Vol. 15 *Molecular Electronics: An Introduction to Theory and Experiment*
Second Edition
by Juan Carlos Cuevas (Universidad Autónoma de Madrid, Spain)
and Elke Scheer (Universität Konstanz, Germany)

Vol. 14 *Synthesis and Applications of Optically Active Nanomaterials*
by Feng Bai (Henan University, China) and
Hongyou Fan (Sandia National Laboratories, USA)

Vol. 13 *Nanoelectronics: A Molecular View*
by Avik Ghosh (University of Virginia, USA)

Vol. 12 *Nanomaterials for Photocatalytic Chemistry*
edited by Yugang Sun (Temple University, USA)

Vol. 11 *Molecular Bioelectronics: The 19 Years of Progress*
Second Edition
by Nicolini Claudio (University of Genoa, Italy)

*For the complete list of volumes in this series, please visit
www.worldscientific.com/series/wssnn

Printed in the United States
By Bookmasters